T0329562

Mechanical Properties and Processing of Ceramic Binary, Ternary, and Composite Systems

Mechanical Properties and Processing of Ceramic Binary, Ternary, and Composite Systems

A Collection of Papers Presented at the 32nd International Conference on Advanced Ceramics and Composites January 27–February 1, 2008 Daytona Beach, Florida

Editors

Jonathan Salem

Greg Hilmas

William Fahrenholtz

Volume Editors

Tatsuki Ohji

Andrew Wereszczak

A John Wiley & Sons, Inc., Publication

Published by John Wiley & Sons, Inc., Hoboken, New Jersey.
Published simultaneously in Canada.

For general information on our other products and services or for technical support, please contact our
Customer Care Department within the United States at (800) 762-2974, outside the United States at
(317) 572-3993 or fax (317) 572-4002.

Wiley also publishes its books in a variety of electronic formats. Some content that appears in print may
not be available in electronic format. For information about Wiley products, visit our web site at
www.wiley.com.

Library of Congress Cataloging-in-Publication Data is available.

ISBN 978-0-470-34492-7

10 9 8 7 6 5 4 3 2 1

Contents

SILICON CARBIDE, CARBON AND OXIDE BASED COMPOSITES

Preface

This volume contains papers presented in the Mechanical Behavior and Structural Design of Monolithic and Composite Ceramics symposium of the 32nd International Conference & Exposition on Advanced Ceramics & Composites held on January 27–February 1, 2008 at Daytona Beach, Florida.

This volume emphasizes the processing and properties of binary, ternary and composite systems, which are being developed to broaden the application of ceramic in structural systems. It is a continuation in the development of a forum for the discussion of recent developments and applications of binary, ternary, and composite systems.

The papers presented at the symposium represented research from 20 countries and demonstrate the worldwide interest in the processing and properties of complex ceramic systems. The organization of the symposium and the publication of this proceeding were possible thanks to the professional staff of The American Ceramic Society and the tireless dedication of many Engineering Ceramics Division members. We would especially like to express our sincere thanks to the symposia organizers, session chairs, presenters and conference attendees, for their efforts and enthusiastic participation in the vibrant and cutting-edge symposium.

Jonathan Salem
NASA Glenn Research Center

Greg Hilmas
Missouri University of Science and Technology

William Fahrenholtz
Missouri University of Science and Technology

Introduction

Organized by the Engineering Ceramics Division (ECD) in conjunction with the Basic Science Division (BSD) of The American Ceramic Society (ACerS), the 32nd International Conference on Advanced Ceramics and Composites (ICACC) was held on January 27 to February 1, 2008, in Daytona Beach, Florida. 2008 was the second year that the meeting venue changed from Cocoa Beach, where ICACC was originated in January 1977 and was fostered to establish a meeting that is today the most preeminent international conference on advanced ceramics and composites

The 32nd ICACC hosted 1,247 attendees from 40 countries and 724 presentations on topics ranging from ceramic nanomaterials to structural reliability of ceramic components, demonstrating the linkage between materials science developments at the atomic level and macro level structural applications. The conference was organized into the following symposia and focused sessions:

Symposium 1	Mechanical Behavior and Structural Design of Monolithic and Composite Ceramics
Symposium 2	Advanced Ceramic Coatings for Structural, Environmental, and Functional Applications
Symposium 3	5th International Symposium on Solid Oxide Fuel Cells (SOFC): Materials, Science, and Technology
Symposium 4	Ceramic Armor
Symposium 5	Next Generation Bioceramics
Symposium 6	2nd International Symposium on Thermoelectric Materials for Power Conversion Applications
Symposium 7	2nd International Symposium on Nanostructured Materials and Nanotechnology: Development and Applications
Symposium 8	Advanced Processing & Manufacturing Technologies for Structural & Multifunctional Materials and Systems (APMT): An International Symposium in Honor of Prof. Yoshinari Miyamoto
Symposium 9	Porous Ceramics: Novel Developments and Applications

Symposium 10	Basic Science of Multifunctional Ceramics
Symposium 11	Science of Ceramic Interfaces: An International Symposium Memorializing Dr. Rowland M. Cannon
Focused Session 1	Geopolymers
Focused Session 2	Materials for Solid State Lighting

Peer reviewed papers were divided into nine issues of the 2008 Ceramic Engineering & Science Proceedings (CESP); Volume 29, Issues 2-10, as outlined below:

- Mechanical Properties and Processing of Ceramic Binary, Ternary and Composite Systems, Vol. 29, Is 2 (includes papers from symposium 1)
- Corrosion, Wear, Fatigue, and Reliability of Ceramics, Vol. 29, Is 3 (includes papers from symposium 1)
- Advanced Ceramic Coatings and Interfaces III, Vol. 29, Is 4 (includes papers from symposium 2)
- Advances in Solid Oxide Fuel Cells IV, Vol. 29, Is 5 (includes papers from symposium 3)
- Advances in Ceramic Armor IV, Vol. 29, Is 6 (includes papers from symposium 4)
- Advances in Bioceramics and Porous Ceramics, Vol. 29, Is 7 (includes papers from symposia 5 and 9)
- Nanostructured Materials and Nanotechnology II, Vol. 29, Is 8 (includes papers from symposium 7)
- Advanced Processing and Manufacturing Technologies for Structural and Multifunctional Materials II, Vol. 29, Is 9 (includes papers from symposium 8)
- Developments in Strategic Materials, Vol. 29, Is 10 (includes papers from symposia 6, 10, and 11, and focused sessions 1 and 2)

The organization of the Daytona Beach meeting and the publication of these proceedings were possible thanks to the professional staff of ACerS and the tireless dedication of many ECD and BSD members. We would especially like to express our sincere thanks to the symposia organizers, session chairs, presenters and conference attendees, for their efforts and enthusiastic participation in the vibrant and cutting-edge conference.

ACerS and the ECD invite you to attend the 33rd International Conference on Advanced Ceramics and Composites (http://www.ceramics.org/daytona2009) January 18–23, 2009 in Daytona Beach, Florida.

TATSUKI OHJI and ANDREW A. WERESZCZAK, Volume Editors
July 2008

Binary and Ternary Ceramics

SYNTHESIS AND PHASE DEVELOPMENT IN THE Cr-Al-N SYSTEM

M-L. Antti[1] , Y-B. Cheng[2] and M. Odén[3]
[1]Luleå University of Technology, Division of Engineering Materials, SE 971 87 Luleå, Sweden
[2]Department of Materials Engineering, Monash University, 3800 Victoria, Australia
[3]Division of Nanostructured Materials, Linköping University, SE 581 83, Sweden

ABSTRACT
The ternary nitride system Cr-Al-N has been investigated by sintering different powder compositions. The powder compositions belong to four groups, AlN- + Cr-powder (5 compositions between 20-90 molar% AlN), Al- + Cr_2N-powder (5 compositions between 15-80 molar% Cr_2N), AlN- + Cr_2N-powder (50- and 90 molar% Cr_2N) and Al- + Cr-powder. The powders were dry mixed and pressed into pellets by uniaxial pressing followed by cold isostatic pressing (CIP). Sintering took place in a graphite lined reaction bonding furnace under nitrogen atmosphere at three different temperatures, 1350°C, 1500°C and 1800°C and in an alumina tube furnace in order to avoid access to carbon. Holding times were varied, from 2 hours up to 72 hours.
The phase development was evaluated by thermal analysis and XRD. CrAlN was formed at 1350°C but decomposed at higher temperatures. Both pure Al and Cr-powder were prone to react with carbon in the graphite furnace. Thermal analysis showed a sublimation of Cr_2N at temperatures around 1050°C and nitridation of pure Al-powder between 680-750°C and of pure Cr-powder between 610-1080°C. Samples with pure Al-powder showed a very large expansion due to melting of aluminium in combination with nitridation. AlN was found to be more stable than Cr_2N at higher temperatures and longer holding times. The mixtures of Al-+Cr-powder produced an intermediate Al-Cr-phase.

INTRODUCTION

Chromium aluminium nitride has shown promising properties for cutting and wear applications, such as high hardness, wear- and oxidation resistance[1,2]. Increasing the amount of aluminium in Cr-N coatings increases the oxidation resistance of the film by formation of an aluminium oxide layer on the surface[3]. There are many reports on thin film production of CrAlN-films[4,5,6], but the material is much less studied in bulk form.

CrAlN crystallizes in two different ways depending on the AlN content. The crystal structure is cubic B1 NaCl-structure CrN for lower amount of AlN and hexagonal B4 wurtzite-structure AlN (w-AlN) for higher amount of AlN[7].

Little work has been reported on bulk preparation of CrAlN. The aim of this work is to investigate the phase development in the Cr-Al-N system during sintering in different temperatures.

MATERIAL

The material in this study consisted of aluminium nitride and pure chromium powder from Alfa Aesar (Johnson Matthey, Karlsruhe, Germany), chromium(III)nitride and aluminium powder from Sigma-Aldrich (Munich, Germany). The particle sizes of the powders were around 40 microns. The chromium(III)nitride powder consists of 85% Cr_2N and 15% CrN. Four different groups of compositions were made, as shown in figure 1.

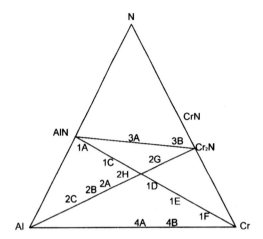

Figure 1. Compositional distribution of the tested samples.

Group 1 consists of 5 compositions of a mixture of AlN-powder and pure Cr-powder. Group 2 consists of 5 compositions of a Al- and Cr_2N -powder mixture. Group 3 is two different mixtures of AlN and Cr_2N -powders. Group 4 consists of two different compositions of Al- and Cr-powders. Table I shows the molar compositions of the different samples.

Table I. Compositions of all samples, in molar fractions.

Sample ID	AlN	Cr	Sample ID	Cr_2N	Al
1A	0.87	0.13	2A	0.49	0.51
1C	0.65	0.35	2B	0.36	0.64
1D	0.50	0.50	2C	0.25	0.75
1E	0.37	0.63	2G	0.79	0.21
1F	0.20	0.80	2H	0.64	0.36
3A	0.26			0.74	
3B	0.04			0.96	
4A		0.5			0.5
4B		0.67			0.33

EXPERIMENTAL

The powders were dry mixed and milled in a mortar and then compacted into pellets with a diameter of 10 mm by uniaxial pressing (MFL Systeme Prüf und Mess:UPD6, Mannheim, Germany) at 10 MPa, followed by cold isostatic pressing (Autoclave Engineers: STD, Erie, PA, USA) at 200 MPa. The samples were sintered in an alumina lined tube furnace (Hereaus, Hanau, Germany) at 1350°C under flowing nitrogen atmosphere with a holding time of 2 hours and in a reaction sintering furnace with graphite heating elements (Conrad Engelke Technik, Hannover, Germany) at 1350, 1500 and

1800°C in a nitrogen atmosphere during 4 hours. In the graphite lined furnace the samples were embedded in boron nitride. Some of the samples were sintered at lower temperatures in the alumina lined furnace, at 750 and 850°C respectively.

Heat treated samples were investigated with X-ray diffraction (XRD) (Philips:MRD, CuKα radiation and proportional detector, Almelo, The Netherlands). The X-ray diffractograms were recorded between 10 and 100° 2θ. The phase fractions were determined by fitting a Gaussian function to the three main XRD peaks of each phase present. The relative intensity ratio, Ix/It, where Ix is the sum of the intensity of the three main peaks of phase x and It is the total intensity, i.e. the sum of the intensity of the three main peaks of all the constituent phases, is then a measure of the phase fraction of phase x.

Thermal analyses were performed in a dilatometer (Netzsch: DIL 402 C, Selb, Germany) and in a differential scanning calorimeter (DSC Netzsch: STA 449 C Jupiter, Selb, Germany) under flowing nitrogen atmosphere up to 1550°C with a heating rate of 10°C/min and cooling rate 20°C/min.

RESULTS AND DISCUSSION

Sintering at 1350°C for 2 hours gave the largest amount of chromium aluminium nitride. Almost all samples contained the phases AlN, Cr_2N and Cr_xAl_yN at different amounts. Group 2 contained more Cr_xAl_yN than group 1, suggesting that pure aluminium more easily reacts with Cr_2N, than chromium does with AlN. The compositions 2A and 2H contain the largest amount of Cr_xAl_yN, around 40 vol%. Resulting XRD graphs for the compositions in group 2 sintered at 1350°C are shown in figure 2.

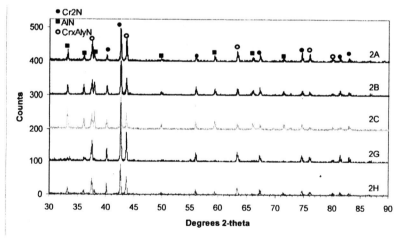

Figure 2. The samples of group 2 sintered at 1350°C.

Sample 2A was also sintered at 750° and 850°C, and a comparison of the result for three different sintering temperatures is shown in figure 3. It can be seen that there is more AlN at higher temperatures. The fact that AlN is visible only at higher sintering temperatures even though nitridation of aluminium starts below 700°C, indicates that it is a decomposition phase of the Cr_xAl_yN -phase rather than a nitridation of pure aluminium powder. There is probably a maximum in Cr_xAl_yN content at a temperature lower than 1350°C but higher than 850°C and this will be investigated further.

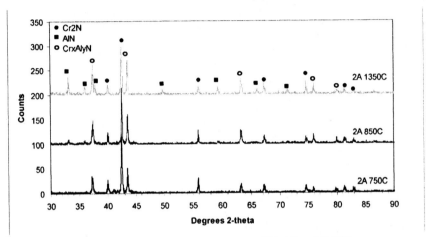

Figure 3. Sample 2A sintered at three different temperatures, 750, 850 and 1350°C.

Group 1 consisted generally of less Cr_xAl_yN -phase than group 2. The Cr_xAl_yN phase also developed at a higher temperature in group 1. As shown for sample 2A in figure 3 above, there is Cr_xAl_yN already at 750°C, but for sample 1D, there is only AlN and pure Cr even at 850°C, at a holding time of 2 hours, see figure 4 below.

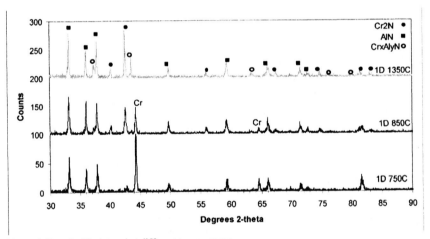

Figure 4. Sample 1D sintered at different temperatures.

The samples of group 4, i.e. a mixture of chromium and aluminium, were sintered in 1350°C, 850°C and 750°C and no Cr_xAl_yN phase was detected. Group 4 samples showed an intermediate Al-Cr phase for all temperatures. However, exact phase composition could not be determined. The resulting XRD-results of sample 4B sintered at 750°C, 850°C and 1350°C are shown in figure 5, with the intermediate Al-Cr phase denoted δ. At 750 and 850°C the sample contains only the Al-Cr-phase and pure Cr. At 1350°C the chromium has been nitridated into Cr_2N and AlN is now visible as a decomposition phase of the Al-Cr-phase. The Al-Cr-phase could not be nitridated into a Cr_xAl_yN-phase even at 1350°C. The same behaviour was detected for sample 4A.

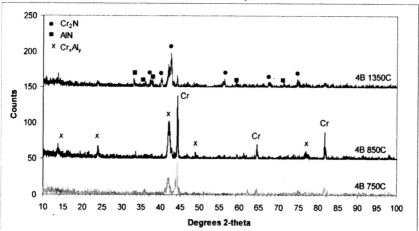

Figure 5. X-ray diffractograms of sample 4B sintered at different temperatures.

The results of the curve fitting showed that there is no shift in the Cr_2N peaks with respect to aluminium content, which indicates that there is no solid solution between Al and Cr_2N. Figure 6 shows the influence of the total amount of aluminium in the sample, i.e aluminium- and aluminium nitride on the 200 peak position of the Cr_xAl_yN phase. The amount of each phase in the samples sintered at 1350°C in the alumina lined furnace and calculated from the intensity relationship is shown in table II and visualised in figure 7.

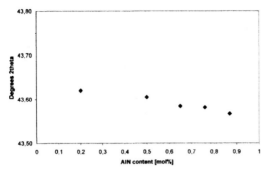

a) Group 1

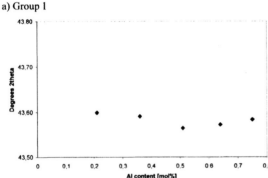

b) Group 2

Figure 6. Peak position of main peak of Cr_xAl_yN phase as a function of total aluminium content (molar%) in the sample for (a) group 1 and (b) group 2.

Table II. Amount of phases in vol% based on intensity relationships of 3 main peaks.

Sample	Cr_xAl_yN (vol%)	Cr_2N (vol%)	AlN (vol%)
1A	7	10	83
1C	15	21	64
1D	15	32	53
1E	10	47	43
1F	16	67	17
2A	41	38	21
2B	31	44	25
2C	25	40	36
2G	37	49	14
2H	42	43	15
3A	21	43	36
3B	0	100	0

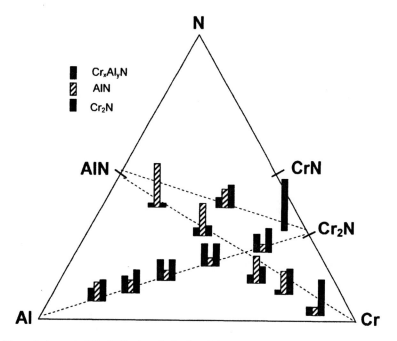

Figure 7. Amount of Cr_xAl_yN -phase in the Cr-Al-N system at 1350°C.

Of the temperatures tested in this study, 1350°C was the temperature that gave the largest amount of Cr_xAl_yN. Longer holding time at 1350°C or sintering at 1500°C and 1800°C lead to a decomposition of the Cr_xAl_yN phase into AlN and Cr_2N. In addition, a reaction occurred between the boron nitride powder that was used for embedding samples in the graphite lined furnace and pure chromium powder starting at a temperature of 1500°C, depending strongly on temperature and holding time. After 72 hours at 1500°C the only phase present was Cr(Al)B, i.e. a CrB phase with Al as a substitutional solid solution. The reaction with BN was naturally more prominent in group 1, consisting of pure Cr as a contrast to group 2 that contains Cr_2N from the start. Chromium was also easily reacting with carbon in the graphite lined furnace. The higher the temperature the more AlN and less Cr_2N, indicating that AlN is a more stable phase at higher temperatures.

The results of the dilatometer runs for sample 1D, 2A and 3A are shown in figure 8. They are representative for all the samples in the different groups. For sample 2A there is a sudden and large increase in length at 670°C which is due to the melting and nitridation of aluminium. Samples 1D and 3A has a much smaller expansion and over a larger temperature range.

The results of the DSC/TG runs for sample 1D and 2A are shown in figure 9. Also here the samples in the same group follow the same trend in behaviour. For group 1 there is a large mass increase between 610-1080°C due to nitridation of chromium. Between 1080 and1150°C there is a decrease in mass, due to sublimation of chromium nitride. Group 2 shows a similar behaviour as group 1. Sample 2A shows a large mass increase between 680°C and 750°C and in the same temperature interval the DSC curve shows an exothermic reaction. This is probably due to nitridation of pure aluminium. The nitridation can start at slightly different temperatures depending on the nitrogen pressure[8]. Aluminium is melting due to the heat of reaction generated during nitridation. A nitride layer grows on the surface of the pellet. This nitride layer prevents the nitrogen from getting in contact with the inner part of the pellet. The temperature of the surrounding nitrogen atmosphere is lower than the melting point of aluminium and the molten aluminium will solidify and crack the nitride layer on the surface, letting the nitrogen into the inner parts of the pellet and the process of melting will start again[8]. Between 1070 and 1170°C there is a mass loss for all samples in group 2. The mass loss is larger the more Cr$_2$N the sample contains together with an endothermic reaction indicates a sublimation of chromium nitride.

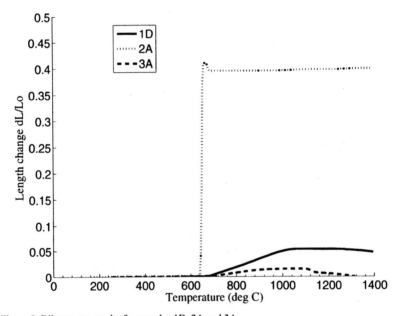

Figure 8. Dilatometer results for samples 1D, 2A and 3A.

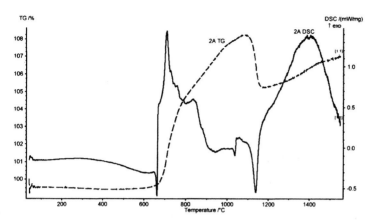

a) Sample 2A

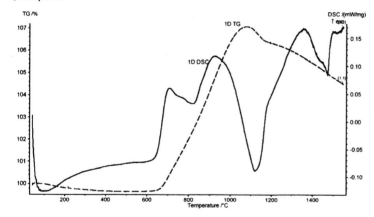

b) Sample 1D

Figure 9. DSC/TG results for samples 1D and 2A.

CONCLUSIONS

- The largest amount of Cr_xAl_yN -phase was found at a temperature of 1350°C and 2 hours holding time in this study.
- The Cr_xAl_yN -phase and the Al-Cr-phase have started to decompose at 1350°C into AlN and Cr_2N indicating that there is an optimum temperature for Cr_xAl_yN reaction below 1350°C.
- In group 4 (i.e. a powder mixture of Al + Cr) there is an intermediate phase of Al-Cr instead of Cr_xAl_yN –phase.
- The Al-Cr-phase has not been nitridated even at 1350°C.

- AlN is more stable than Cr_2N at higher temperatures and longer holding times.
- The largest amount of Cr_xAl_yN –phase was found in group 2 (i.e. a powder mixture of Al + Cr_2N)

REFERENCES

1. M. Kawate, A.K. Hashimoto and Suzuki, Oxidation resistance of Cr1-xAlxN and Ti1-xAlxN films, T. *Surface & Coatings Technology.* **165** (2),163-167, (2003)
2. Y. Makino and K. Nogi, Synthesis of pseudobinary Cr-Al-N films with B1 sructure by rf-assisted magnetron sputtering method, *Surface & Coatings Technology.* **98** (1-3),1008-1012, (1998)
3. I.W. Park et al, Microstructures, mechanical properties, and tribological behaviors of Cr–Al–N, Cr–Si–N, and Cr–Al–Si–N coatings by a hybrid coating system, *Surface & Coatings Technology* **210** (9-11), 5223-5227, (2007)
4. J. Lin, B. Mishra, J.J. Moore and W.D Sproul, Microstructure, mkechanical and tribological properties of Cr1-xAlxN films deposited by pulsed-closed field unbalanced magnetron sputtering (P-CFUBMS), *Surface & Coatings Technology* **201**, 4329-4334,(2006)
5. J. Romero et. al, CrAlN coatings deposited by cathodic arc evaporation at different substrate bias, *Thin Solid Films* **515** ,113-117, (2006)
6. H. Willmann et al, Thermal stability of Al-Cr-N hard coatings, *Scripta Materialia* **54** 1847-1851, (2006)
7. P.H Mayrhofer, H. Willmann and A.E Reiter, Structure Evolution of Cr-Al-N Hard Coatings, *Society of vacuum Coaters*, 505/856-7188, ISSN 0737-5921,(2206)
8. Okada, T. et al., Direct nitridation of aluminium compacts at low temperature, *Journal of Materials Science,* **35** (12), 3105-3111(2000)

PHASE EVOLUTION AND PROPERTIES OF Ti₂AlN BASED MATERIALS, OBTAINED BY SHS METHOD

L. Chlubny, J. Lis, M.M. Bućko
University of Science and Technology, Faculty of Material Engineering and Ceramics, Department of Technology of Ceramics and Refractories
Al. Mickiewicza 30, 30-059, Cracow, Poland

ABSTRACT

Ternary compounds in the Ti-Al-N system were synthesized by self-propagating high temperature synthesis (SHS). Using elemental precursors, SHS produced a mixture containing mainly TiN and AlN. However, when TiAl was used as a precursor, a significant amount of Ti_2AlN was formed. The powders were densified by subsequent hot pressing at temperatures ranging from 1100°C to 1300°C. Hot pressing at 1300°C or 1400°C resulted in near complete conversion to the desired Ti_2AlN phase, while higher temperatures resulted in the decomposition of the ternary compound to TiN, TiC, and AlN. The Young's modulus, shear modulus, hardness and fracture toughness of the composites were measured. Hardness reached a maximum value of ~7 GPa after hot pressing at 1200°C. For materials densified at 1300°C or higher, toughness could not be measured using the direct crack method due to pseudo-plastic deformation.

1. INTRODUCTION

Among many covalent materials such as carbides or nitrides there is a group of ternary compounds referred in literature as H-phases, Hägg-phases, Novotny-phases or thermodynamically stable nanolaminates. These compounds have a $M_{n+1}AX_n$ stoichiometry, where M is an early transition metal, A is an element of A groups (mostly IIIA or IVA) and X is carbon and/or nitrogen. Heterodesmic structures of these phases are hexagonal, P6₃/mmc, and specifically layered. They consist of alternate near close-packed layers of M_6X octahedrons with strong covalent bonds and layers of A atoms located at the centre of trigonal prisms. The M_6X octahedra, similar to those forming respective binary carbides, are connected one to another by shared edges. Variability of chemical composition of the nanolaminates is usually labelled by the symbol describing their stoichiometry, e.g. Ti_2AlN represents 211 type phase and Ti_3AlC_2 – 312 typ. Structurally, differences between the respective phases consist in the number of M layers separating the A-layers: in the 211's there are two whereas in the 321's three M-layers [1-3]. The layered, heterodesmic structure of MAX phases led to an extraordinary set of properties. These materials combine properties of ceramics like high stiffness, moderately low coefficient of thermal expansion and excellent thermal and chemical resistance with low hardness, good compressive strength, high fracture toughness, ductile behavior, good electrical and thermal conductivity characteristic for metals. They can be used to produce ceramic armor based on functionally graded materials (FGM) or as a matrix in ceramic-based composites reinforced by covalent phases.

The objective of this work was examine of phase evolution of Ti-Al-N materials during hot pressing of the SHS-derived powders in the temperature range of 1100-1600°C.

2. PREPARATION

Intermetallic materials in the Ti-Al system were used as a precursors for synthesis of Ti_2AlN powders. In the first stage of the experiment TiAl powder was synthesized by SHS method [4]. Titanium hydride powder, TiH_2, and metallic aluminium powder with grain sizes below 10 μm were used as sources of titanium and aluminium. The mixture for SHS had a molar ratio of 1:1. The powders were mixed in dry isopropanol using a ball-mill. The dried powder was placed in a graphite crucible which was heated in a graphite furnace in the argon atmosphere up to 1200°C when SHS reaction was initiated. The product was initially crushed in a roll crusher to the grain size ca. 1 mm and then the powders were ground in the rotary-vibratory mill for 8 hours in isopropanol, using WC balls as a grinding medium, to the grain size ca. 10μm.

The synthesis of Ti$_2$AlN was conducted by SHS with a local ignition system and different precursors. The first set of precursors was a mixture of titanium and aluminium powders set in appropriate stoichiometric ratio ignited in a nitrogen the atmosphere. The expected chemical reaction was the direct synthesis of anticipated compound:

$$4\,Ti + 2\,Al + N_2 \rightarrow 2\,Ti_2AlN \qquad (1)$$

The intermetallic compound (TiAl) that was synthesized by SHS was used in the second synthesis. It was mixed in an appropriate ratio with other precursors:

$$2\,TiAl + 2\,Ti + N_2 \rightarrow 2\,Ti_2AlN \qquad (2)$$

Homogenized mixtures were placed in high-pressure reactor as a loose bed in a graphite holder. The SHS synthesis was initiated by a local ignition and performed at 0.5 MPa of nitrogen.

SHS derived powders with the highest content of Ti$_2$AlN was chosen for the hot-pressing process, based on XRD phase analysis. Selected powder was hot-pressed in the temperature range of 1100-1600°C in a constant nitrogen flow. The annealing time at the maximum temperature was 1 hour. Graphite mold of 1 inch diameter was used for hot pressing process [4].

The X-ray diffraction analysis method was applied to determine phase composition of the synthesised materials. The basis of phase analysis were data from ICCD [5]. Phase quantities were calculated by comparison method, based on knowledge of relative intensity ratios (RIR) [6].

Some mechanical properties of hot pressed dense bodies such as Young's modulus, shear modulus, Vickers' hardness and fracture toughness were examined. For examination of elastic properties ultrasonic method was applied. For Vickers' hardness and fracture toughness' indentation was used on seven samples of each material.

3. Results and Discussion

X-ray diffraction analysis proved that TiAl synthesised by SHS was almost phase pure and contained only about 5% of Ti$_3$Al impurities (Figure 1) [7].

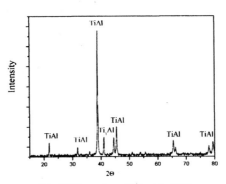

Figure1. XRD pattern of the TiAl powders obtained by SHS

Ti$_2$AlN synthesized by the reaction between the mixture of elements did not show positive results. The dominant phases were titanium nitride (TiN) and hexagonal aluminium nitride AlN. Some amount of non-reacted intermetallic phases were also found in the product.

Better results were achieved in the case of the SHS synthesis using a homogenized mixture of TiAl, metallic titanium powder and nitrogen. The dominant phase was Ti$_2$AlN accompanied by TiN and Ti$_3$AlN and Ti$_3$Al. The phase compositions of each product are presented in Table I [7].

Table I. Phase composition in Ti-Al-N system after the SHS synthesis.

Precusor	Composition, % wt.
4 Ti + 2 Al + N$_2$	66% TiN, 15% AlN, 7% Ti$_3$AlN, 12% (TiAl+Ti$_3$Al+AlTi$_3$)
2 TiAl + 2 Ti + N$_2$	57% Ti$_2$AlN, 24% TiN, 11% Ti$_3$Al, 8% Ti$_3$AlN

Table II. Sintering conditions

Material	Temperature [°C]	Time [h]	Atmosphere
2 TiAl + 2 Ti + N$_2$	1100	0,5	Nitrogen
↓	1100	1	Nitrogen
After SHS	1200	1	Nitrogen
57 % Ti$_2$AlN,	1300	1	Nitrogen
24 % TiN, 11	1400	1	Nitrogen
%Ti$_3$Al, 8 % Ti$_3$AlN	1600	1	Nitrogen

Sintering conditions are presented in the Table II [7]. XRD patterns of hot pressed materials are presented on Figure 2 and Figure 3 [7].

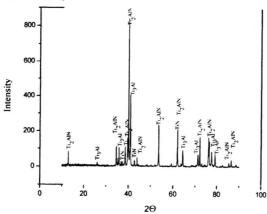

Figure 2 XRD pattern of Ti$_2$AlN sintered at 1100°C

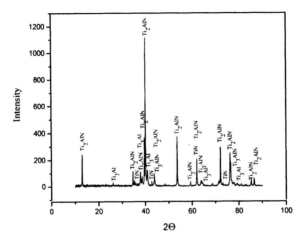

Figure 3 XRD pattern of Ti₂AlN sintered at 1300°C

It was observed that a chemical reaction took place in the samples hot pressed in nitrogen, so this process could be called reaction sintering. The powder before sintering contained four compounds with Ti₂AlN as the major phase (~57%) and other minor phases which were TiN, Ti₃Al and Ti₃AlN. Increasing the sintering temperature from 1100°C to 1300°C led to formation of significant amounts (up to 96%) of the layered compound (Ti₂AlN), accompanied by decreased amounts of Ti₃Al and TiN. At temperatures above 1300°C, Ti₂AlN decomposed completely, resulting in the formation of TiN, TiC, AlN and Ti₃AlC₂. Changes in the phase composition during sintering are summarized in Table III [7].

Table III. Comparison of phase compositions of sintered materials

Material (phase composition before sintering, Sintering temperture, time, atmosphere)		Phase composition
57 % Ti₂AlN, 24 % TiN, 11 % Ti₃Al, 8 % Ti₃AlN	1100°C, 0,5h, nitrogen	76% Ti₂AlN, 16% Ti₃Al , 4 % TiN, 4% TiAl₃
	1100°C, 1h, nitrogen	80% Ti₂AlN, 16% Ti₃Al , 4 % TiN
	1200°C, 1h, nitrogen	90% Ti₂AlN, 8% Ti₃Al , 2 % TiN
	1300°C, 1h, nitrogen	96% Ti₂AlN, 2% Ti₃Al , 1 % TiN, 1% Ti₃AlN
	1400°C, 1h, nitrogen	100% Ti₃Al₂N₂
	1600°C, 1h, nitrogen	36%TiN, 34% TiC, 26% AlN, 3% Ti₃AlC₂

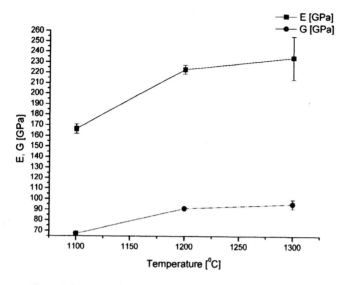

Figure 4 Sintering temperature dependence of Young's and shear moduli

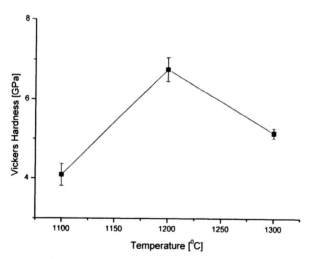

Figure 5 Sintering temperature dependence of Vickers' hardness

In the case of Young's modulus and shear modulus of Ti$_2$AlN materials examined by ultrasonic method, it was observed that increasing the hot-pressing temperature led to an increase in both moduli which is related to the amount of ternary phase in the material. Figure 4 shows changes of the Young's and shear moduli for Ti$_2$AlN material prepared from SHS-derived powder (TiAl and Ti mixture) and hot-pressed at different temperatures. Young's modulus increased with increasing hot pressing temperature to a maximum of ~235 GPa for material hot pressed at 1300°C.

Sintering temperature dependence was also observed for the Vickers' hardness, where the highest values were achieved at 1200°C. The sample with the highest percentage of ternary material (sintered at 1200°C) showed the highest hardness, while samples sintered at 1100°C had lower hardness values. Figure 5 shows the changes in Vickers' hardness for samples sintered at different temperatures.

In the case of fracture toughness, measured by direct crack method, values are presented in Table IV [7]. Increasing the amount of the TiN-type phase, which is much stiffer and much tougher than the ternary phase, caused an increase in the K$_{1c}$ value. It is worth mentioning, pseudo-plastic properties similar the pure nanolaminate phases were observed for the sample sintered at 1300°C. No radial cracks or other changes were observed in the material around the indentation mark, which was consistent with pseudo-plastic behavior. Thus, values of fracture toughness were not measurable the direct crack method for materials densified at 1300°C. Figure 6 micrograph of a Vickers' indentation made on the polished surface of Ti$_2$AlN hot-pressed at 1300°C is presented. Note that no cracking was observed around the indent.

Table IV. Fracture toughness of hot-pressed composites.

Temperature, C	K$_{1c}$ [Pa*m$^{0.5}$]
1100	7.00 ± 0.4
1200	8.2 ± 0.1
1300	not measurable

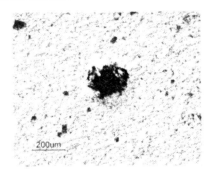

Figure 6 Vickers' indent in Ti$_2$AlN sintered at 1300°C

4. CONCLUSIONS

Ti$_2$AlN powders obtained by selfpropagating high-temperature synthesis were able to be densified by hot pressing. Chemical reactions occurred during the hot-pressing process. Increasing the hot pressing temperature increased the amount of Ti$_2$AlN (up to 96%wt.). Material obtained at 1300°C was nearly single phase nanolaminate Ti$_2$AlN, which may find various applications.

The mechanical properties of the hot pressed materials varied with the densification temperature. The moduli increased as hot pressing temperature increased with Young's modulus reaching a maximum of ~235 GPa for a hot pressing temperature of 1300°C. Hardness values showed a maximum of ~7 GPa for hot pressing at 1200°C, but the values decreased for higher or lower hot pressing temperatures. The direct crack method measured fracture toughness values of more than 7 MPa.ml/2 for materials hot pressed at 1100°C and 1200°C. However, pseudo-plastic deformation was observed in the material hot pressed at 1300°C, which meant that its toughness could not be determined using this method.

ACKNOWLEDGEMENTS:

This work was supported by the Polish Ministry of Science and Higher Education under the grant no. N 507 2112 33.

5. REFERENCES

[1] W. Jeitschko, H. Nowotny, F.Benesovsky, Kohlenstoffhaltige ternare Verbindungen (H-Phase). Monatsh. Chem. 94 672-678, (1963).

[2] H. Nowotny, Structurchemie Einiger Verbindungen der Ubergangsmetalle mit den Elementen C, Si, Ge, Sn. Prog. Solid State Chem. 2 27, (1970)

[3] M.W. Barsoum: The MN+1AXN Phases a New Class of Solids; Thermodynamically Stable Nanolaminates- Prog Solid St. Chem. 28, 201-281, (2000)

[4] L. Chlubny, M.M. Bucko, J. Lis "Intermetalics as a precursors in SHS synthesis of the materials in Ti-Al-C-N system" Advances in Science and Technology, 45 1047-1051, (2006)

[5] "Joint Commitee for Powder Diffraction Standards: International Center for Diffraction Data"

[6] F.H.Chung: Quantitative interpretation of X-ray diffraction patterns, I. Matrix-flushing method of quantitative multicomponent analysis. - J.Appl.Cryst., 7, 513 – 519, (1974a)

[7] L. Chlubny: New materials in Ti-Al-C-N system. - PhD Thesis. AGH-University of Science and Technology, Kraków 2006. (in Polish)

SYNTHESIS OF Ti$_3$SiC$_2$ BY REACTION OF TiC AND Si POWDERS

Ida Kero[1*], Marta-Lena Antti[1], and Magnus Odén[2]
[1]Luleå University of Technology, Department of Applied Physics and Mechanical Engineering, Division of Engineering Materials, 971 87 Luleå, Sweden
[2]Linköping University, Division of Nanostructured Materials, 581 83, Sweden
*Corresponding and presenting author: ida.kero@ltu.se

ABSTRACT

The MAX phase Ti$_3$SiC$_2$ has been synthesized from starting powder mixtures which do not include pure titanium. The presence of pure titanium in a powder is problematic because of its oxidizing, and in the form of a finely divided powder, explosive nature. The aim of this study was to evaluate the synthesis of bulk polycrystalline samples of Ti$_3$SiC$_2$ from a starting powder mixture which is more suited for large scale production.

Titanium silicon carbide MAX phase was synthesized by pressureless sintering of ball milled TiC and Si powders of six different compositions. The sintering reactions were evaluated in situ by dilatometer analysis under flowing argon gas. The as-sintered samples were evaluated using mainly x-ray diffraction (XRD) analysis. This study showed that titanium carbide, silicon carbide and titanium disilicide were present as intermediate or secondary phases in the samples.

Our results indicate that TiSi$_2$ is an intermediate phase to the formation of Ti$_3$SiC$_2$ when excess Si is present. The excess of silicon also proved beneficial for the synthesis of the MAX phase and there is a Si content which is optimal with respect to the maximum MAX phase content of the final product. The Ti$_3$SiC$_2$ was found to decompose into TiC and gaseous Si at high temperatures.

INTRODUCTION

Ti$_3$SiC$_2$ is a ceramic material which has received increased attention over the past decade because of its attractive combination of properties. It belongs to a group of ternary layered nitrides and carbides known as the MAX phases; with general formula M$_{n+1}$AX$_n$, where M is an early transition metal, A is an element from groups 12-16 in the periodic table of the elements, X is either nitrogen or carbon and n is an integer 1-3. Titanium silicon carbide is the most well known of the MAX phases and it combines some of the most appreciated qualities of ceramics with those of metals, e.g. it is refractory, light weight and stiff but also damage tolerant, machinable with conventional tools and not susceptible to thermal shock.[1-5]

Ti$_3$SiC$_2$ is most often prepared from starting powder mixtures including pure titanium, such as Ti/Si/C[6-11], Ti/C/SiC[2,12-15] and Ti/Si/TiC[16-18] by use of different powder metallurgical synthesis methods, e.g. hot pressing,[2,8,10] hot isostatic pressing,[12,17] vacuum sintering,[7,11,15,18] pressureless sintering,[6,9,13,14,16] mechanical alloying[19-21] and self-propagation high-temperature synthesis.[22] Pure titanium powder is highly oxidizing and thus has to be handled with great care under strict safety precautions.[23]

Monolithic Ti$_3$SiC$_2$ has been reported by a handful of authors[2,12,16,18] but secondary phases such as titanium carbides, silicon carbides and/or titanium silicides are still common in the final products. Different intermediate phases and reactions have been proposed for the formation of Ti$_3$SiC$_2$.

El-Raghy and Barsoum[12] worked with Ti/C/SiC powders and reported that Ti$_3$SiC$_2$ nucleated and grew within Ti$_5$Si$_3$C$_x$ grains. They proposed a series of reactions where TiC$_{0.5}$ and Ti$_5$Si$_3$C would serve as intermediate phases. Wu et al. initially confirmed[13] the observations of El-Raghy and Barsoum and later added[14] that the reactions took place simultaneously. Istomin et al.[15] proposed in 2006 a different set of reactions where TiC and Si would be the first intermediate phases to form. These would then react to form Ti$_5$Si$_3$ and TiSi$_2$ which would in turn consume more TiC and Si to produce Ti$_3$SiC$_2$.

Li et al.[10] found TiC, Ti$_5$Si$_3$ and TiSi$_2$ in their samples made from elemental starting powders. They stated that the starting powder composition, especially with regards to the silicon content, determined the phase composition of the final product. Yang et al.[18] reported comparable observations to the effect of varying silicon content to Ti/Si/TiC powders and concluded that the optimum silicon content was 10 % excess to the stoichiometric composition. They stated that excess silicon would compensate for Si loss by evaporation whereas too much added Si would promote TiSi$_2$ formation over Ti$_3$SiC$_2$.

Attempts have also been made to synthesize Ti$_3$SiC$_2$ from TiC/Si powder mixtures, thus avoiding the pure titanium. Radhakrishnan et al.[24] reported that a reaction of 3TiC/2Si powders would first generate the intermediate phase TiSi$_2$ at a temperature of 1170 °C. In a second step the silicide would be consumed as Ti$_3$SiC$_2$ and SiC would form. Li et al.[25] published in 2004 a differential scanning calorimetry (DSC) study, suggesting that the reaction between 3TiC/2Si powders would start at 1340 °C. They found SiC and TiC to be present as secondary phases in their final products.

Racault et al.[7] were first to report the decomposition of Ti$_3$SiC$_2$ into TiC$_x$ and gaseous Si. Wu et al.[13] who observed that the presence of titanium carbide in the sample was deleterious to the thermochemical stability of Ti$_3$SiC$_2$. Gao et al.[17] stated that the stability was related to the vapor pressure of silicon in the furnace. Li et al.[25] suggested another reaction formula for the decomposition of Ti$_3$SiC$_2$ by carburization where SiC instead of Si would be produced. We have previously reported observations of MAX phase formation via the intermediate phase of TiSi$_2$ and a decomposition of the MAX phase into TiC and gaseous Si.[26] The aim of this study was to evaluate the influence of excess silicon in the starting powders on the amount of MAX phase obtained in the final products as well as the reaction mechanisms governing its formation.

EXPERIMENTAL

The powders used were TiC (Aldrich, -325 mesh) and Si (Aldrich, -325 mesh); they were mixed at six different ratios. The B composition with the smallest amount of silicon had a TiC/Si ratio of 3:1 and the G composition with the largest amount of silicon had a TiC/Si ratio of 3:2.8. The different starting powder compositions are summarized in Table 1. The powders were mixed and milled in a ball mill and then compacted by uniaxial pressing to 10 MPa, and cold isostatic pressing to 300 MPa. The samples were sintered in a dilatometer with graphite heating elements and sample holder, under flowing argon gas. Sintering was also performed in a furnace with graphite heating elements in which the samples were embedded in boron nitride. Here the sintering took place at a temperature of 1250°C and with 2.5 hour holding time, under a vacuum of approximately 0.02 mbar. The samples were crushed into a powder, and analyzed by x-ray diffractometry (XRD) using Cu radiation and a proportional detector. The phase fractions were determined using the direct comparison method[27]. Here, the integrated intensity for a minimum of three diffraction lines of each phase were summed and the volume fraction of the individual phases was calculated by:

$$V_i = \frac{\frac{1}{n}\sum_{j=1}^{n}\frac{I_i^j}{R_i^j}}{\frac{1}{n}\sum_{j=1}^{n}\frac{I_\alpha^j}{R_\alpha^j}+\frac{1}{n}\sum_{j=1}^{n}\frac{I_\beta^j}{R_\beta^j}+\frac{1}{n}\sum_{j=1}^{n}\frac{I_\gamma^j}{R_\gamma^j}+\dots} \tag{1}$$

Where n is the number of hkl peaks for a given phase, V is the volume fraction, I is the integrated intensity and R is the calculated theoretical intensity. The validity of this procedure has been demonstrated for highly anisotropic, two-phase steels by Dickson[28].

Table 1. Starting powder compositions

Sample name	TiC/Si ratio of the starting powder	Expected phases	Phases present
B	3:1	TiC, Ti$_3$SiC$_2$ & SiC	TiC & Ti$_3$SiC$_2$
C	3:2	TiSi$_2$, Ti$_3$SiC$_2$ & SiC	TiC, TiSi$_2$, Ti$_3$SiC$_2$ & SiC
D	3:2.2	TiSi$_2$, Ti$_3$SiC$_2$ & SiC	TiC, TiSi$_2$, Ti$_3$SiC$_2$ & SiC
E	3:2.4	TiSi$_2$, Ti$_3$SiC$_2$ & SiC	TiC, TiSi$_2$, Ti$_3$SiC$_2$ & SiC
F	3:2.6	TiSi$_2$, Ti$_3$SiC$_2$ & SiC	TiC, TiSi$_2$, Ti$_3$SiC$_2$ & SiC
G	3:2.8	TiSi$_2$, Ti$_3$SiC$_2$ & SiC	TiC, TiSi$_2$, Ti$_3$SiC$_2$ & SiC

RESULTS

Figure 1 shows a typical thermal expansion curve for the samples analyzed in this study. Up to 1142°C the thermal expansion is linear and the rate is constant. At 1142°C the curve increases and 1276°C is the onset of a peak. The peak reaches a maximum at 1457°C, after which the sample shrinks abruptly and considerable amounts of silicon is given off. In order to avoid excessive silicon evaporation and instrument contamination, the heating segment was limited to 1500°C for most samples.

The only sample which differs from this behavior is the B sample with the lowest amount of silicon. It is represented by a dotted line in Figure 1 and it has an earlier onset of the peak. As can be seen from the figure the peak begins with a much less distinct "shoulder" at a temperature of about 1238°C. The curve peaks at 1382°C, which is 75° lower than the other samples.

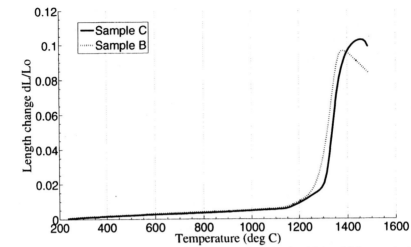

Figure 1: Dilatometer curves of samples B and C with a TiC/Si ratio of 3:1 and 3:2 respectively. The curves of all other samples analyzed followed the same typical pattern as sample C.

Figure 2 is a scanning electron micrograph of the F sample showing a typical microstructure. Backscattered electron mode revealed different chemical compositions of the matrix and the different grains within the matrix. TiC, TiSi$_2$ and SiC phases occur in separate grains distributed in a Ti$_3$SiC$_2$ matrix. A closer look at the TiC grains revealed two types of grains with different morphologies, indicating that TiC is not only a reactant phase but also a product of the decomposition of the MAX phase.[7]

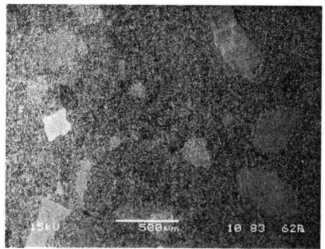

Figure 2: Scanning electron micrograph of the F sample showing a microstructure consisting of grains of TiC, SiC and TiSi$_2$ and a matrix of Ti$_3$SiC$_2$. TiC grains are present in two different morphologies.

Figure 3 shows x-ray diffractograms of the samples sintered under vacuum for 2 hours and 30 min in 1250°C. The B sample with the lowest amount of silicon in the starting powder differs from the other samples. It is the only sample which does not contain silicon carbide (SiC) and titanium disilicide (TiSi$_2$). All samples contained titanium silicon carbide MAX phase (Ti$_3$SiC$_2$) and titanium carbide (TiC).

The relative phase amounts of Ti$_3$SiC$_2$ and TiC are shown for the different starting powders in Figure 4. Most MAX phase was obtained for the 3TiC/2.6Si starting powder composition. The relative amount of TiC is decreasing while the amount of Ti$_3$SiC$_2$ increases, but when the decomposition of Ti$_3$SiC$_2$ causes the MAX phase amount to drop, the TiC content is raised again. This suggests that the TiC is both a reactant, consumed in the MAX phase forming reaction, and a product of the decomposition reaction.

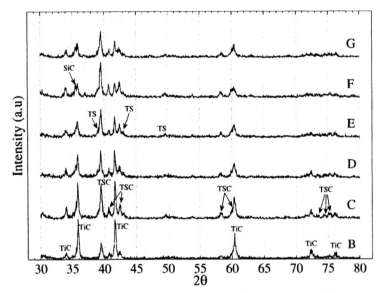

Figure 3: X-ray diffractograms of samples B to G after sintering for 2.5 hours in 1250°C.

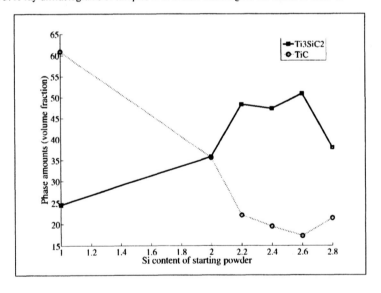

Figure 4: Effect of initial silicon content on the amount of titanium silicon carbide MAX phase (Ti$_3$SiC$_2$) and titanium carbide (TiC) obtained in the final product after sintering for 2.5 hours at 1250°C.

DISCUSSION

The 1200°C isothermal section of the ternary phase equilibrium diagram of the Ti-Si-C system is shown in Figure 5. The composition of sample B is represented in the figure by a circle and the compositions of the other samples are found within the marked rectangle. The B sample had an initial composition within the Ti$_3$SiC$_2$-SiC-TiC three phase area. The sample was found to contain TiC and small amounts of Ti$_3$SiC$_2$. Though the x-ray diffractograms do not indicate any SiC in the samples its presence in small amounts cannot be completely ruled out as the main peaks of SiC can be overlapped by those of the other two phases. The evaporation of silicon has been observed for other samples in this investigation and is commonly reported in the literature[16,18,22,29-33], which may explain the very low amounts of MAX phase and SiC obtained in sample B. As silicon is lost by evaporation, the composition of the samples will shift along the marked line in the figure towards the Ti-C base line of the diagram. Moving towards the TiC corner of the three phase area will deplete the samples of the other two phases.

The initial composition of samples C to G is situated within the TiSi$_2$-SiC-Ti$_3$SiC$_2$ three phase area. These three phases were indeed observed in the samples along with some TiC. The presence of TiC may have more than one reason: it may, of course, be residual, unreacted starting powder or it may be a product of the decomposition of the MAX phase. The two different morphologies observed by SEM suggest that the TiC phase may have more than one origin in which case both of these reasons may be valid. The evaporation of silicon from these samples will shift the phase composition towards the Ti$_3$SiC$_2$-SiC line of the diagram. If the line is traversed due to silicon evaporation TiC will form as a result of an equilibrium phase reaction.

Much more MAX phase was achieved in these samples with excess silicon and most MAX phase was obtained with the 3TiC/2.6Si starting powder. This amount of silicon is likely to compensate for a certain amount of evaporation in such a way that the sample composition is shifted most closely to the Ti$_3$SiC$_2$ phase area.

Samples analyzed in the dilatometer up to temperatures above 1500°C lost considerable amounts of silicon through evaporation that was deposited on the graphite sample holder in the dilatometer. These samples consisted mostly of TiC. Since carbon was present in the furnaces through the graphite heating elements and sample holder/crucible, the decomposition of Ti$_3$SiC$_2$ is likely to follow the reaction initially proposed by Racault et al.[7]:

$$Ti_3SiC_2 + C \rightarrow TiC + Si(g) \qquad (1)$$

As no intermediate phases or unexpected secondary phases were observed in sample B, with no excess silicon, the MAX phase forming reaction is assumed to be a direct displacement reaction as proposed by Radhakrishnan et al.[24]:

$$3TiC + 2Si \rightarrow Ti_3SiC_2 + SiC \qquad (2)$$

The silicide forming reaction of samples C to G is assumed to follow equation 3 as theoretical computations by Li et al.[25] has pointed this reaction out as the thermodynamically most favourable of the plausible reactions.

$$TiC + 3Si \rightarrow TiSi_2 + SiC \qquad (3)$$

As the sample holders and the crucibles used in this study were made of graphite, carbon was readily available during the sintering reactions. All samples with excess silicon had comparable and relatively low amounts of TiSi$_2$. Previous results[34] have shown that the silicide is an intermediate phase to MAX phase formation from these powders. TiSi$_2$ is then likely consumed in a second, Ti$_3$SiC$_2$ forming reaction:

$$3TiSi_2 + 7C \rightarrow Ti_3SiC_2 + SiC \qquad (4)$$

Alternatively, more than one reaction may occur simultaneously in the samples with excess silicon, in which case concurrent reactions according to equations 2 and 3 may compete.

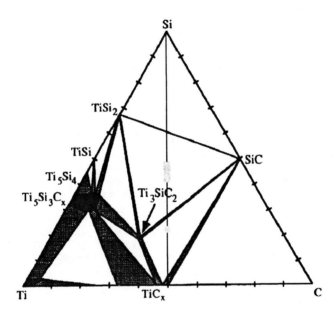

Figure 5: 1200°C isothermal section of the ternary phase equilibrium diagram of the Ti-Si-C system. Illustration by Arunajatesan and Carim[29], original after Ratcliff and Powell[35]. The starting powder compositions used in this study are situated along the marked line; sample B at the marked circle and samples C to G within the marked rectangle.

CONCLUSIONS

- Excess silicon is beneficial for the production of Ti$_3$SiC$_2$. The largest amount of the MAX phase was achieved in the samples with a TiC/Si ratio of 3:2.6, i.e. about 46% Si. The excess silicon is likely to compensate for losses due to evaporation.
- In the samples with excess silicon, TiSi$_2$ and SiC were observed. These are secondary equilibrium phases but the silicide may also take part in a Ti$_3$SiC$_2$ forming reaction.
- In the samples without excess silicon, Ti$_3$SiC$_2$ is assumed to form via a direct solid state displacement reaction.
- The MAX phase decomposes into TiC and gaseous Si at temperatures above 1500°C.

ACKNOWLEDGEMENT

I. Kero gratefully acknowledges the financial support from the Swedish National Graduate School of Space Technology

REFERENCES

[1]Michel W. Barsoum and Tamer El-Raghy, The MAX Phases: Unique New Carbide and Nitride Materials, *American Scientist* **89** (2001).
[2]Michel W. Barsoum and Tamer El-Raghy, Synthesis and Characterization of a Remarkable Ceramic: Ti3SiC2, *J. Am. Ceram. Soc.* **79** (7), 1953 (1996).
[3]Michel W. Barsoum, The Mn+1AXn Phases: A New Class of Solids; Thermodynamically Stable Nanolaminates *Prog. Solid St Chem.* **28**, 201 (2000).
[4]Tamer El-Raghy, Antonios Zavaliangos, Michel W. Barsoum et al., Damage Mechanisms Around Hardness Indentations in Ti3SiC2, *J. Am. Ceram. Soc.* **80**, 513 (1997).
[5]Tamer El-Raghy, Michel W. Barsoum, Antonios Zavaliangos et al., Processing and Mechanical Properties of Ti3SiC2: II, Effect of Grain Size and Deformation Temperature, *J. Am. Ceram. Soc.* **82** (10), 2855 (1999).
[6]R. Radhakrishnan, J. J. Williams, and M. Akinc, Synthesis and high-temperature stability of Ti3siC2, *Journal of Alloys and Compounds* **285**, 85 (199).
[7]C. Racault, F. Langlais, and R. Naslain, Solid-state synthesis and characterization of the ternary phase Ti3SiC2, *Journal of Materials Science* **29**, 3384 (1994).
[8]Yanchun Zhou and Zhimei Sun, Temperature fluctuation/hot pressing synthesis of Ti$_3$SiC$_2$, *Journal of Materials Science* **35**, 4343 (2000).
[9]Zhe Feng Zhang, Zheng Ming Sun, Hitoshi Hashimoto et al., Effects of sintering temperature and Si content on the purity of Ti$_3$SiC$_2$ synthesized from Ti/Si/TiC powders, *Journal of Alloys and Compounds* **352**, 283 (2003).
[10]Shi-Bo Li, Jian-Xin Xie, Li-Tong Zhang et al., Synthesis and some properties of Ti3SiC2 by hot pressing of titanium, silicon and carbon powders, Part 1 - Effect of starting composition on formation of Ti3SiC2 and observation of Ti3SiC2 crystal morphology, *Materials Science and Technology* **19** (10), 1442 (2003).
[11]H. Li, L. M. Peng, M. Gong et al., Preparation and characterization fo Ti3SiC2 powder, *Ceramics International* **30**, 2289 (2004).
[12]Tamer El-Raghy and Michel W. Barsoum, Processing and Mechanical Properties of Ti3SiC2: I, Reaction Path and Microstructure Evolution, *J. Am. Ceram. Soc.* **82** (10), 2849 (1999).
[13]Erdong Wu, Erich H. Kisi, Shane J. Kennedy et al., In Situ Neutron Powder Diffraction Study of Ti$_3$SiC$_2$ Synthesis, *J.Am.Ceram.Soc.* **84** (10), 2281 (2001).

[14]Erdong Wu, Erich H. Kisi, Daniel P. Riley et al., Intermediate Phases in Ti_3SiC_2 Synthesis from Ti/SiC/C Mixtures Studied by Time-Resolved Neutron Diffraction, *J.Am.Ceram.Soc.* **85** (12), 3084 (2002).

[15]P. V. Istomin, A. V. Nadutkin, I. Ryabkov et al., Preparation of Ti_3SiC_2, *Inorganic Materials* **42** (3), 250 (2006).

[16]J. T. Li and Y. Miyamoto, Fabrication of Monolithic Ti_3SiC_2 Ceramic Through Reactive Sintering of Ti/Si/2TiC, *Journal of Materials Synthesis and Processing* **7** (2) (1999).

[17]N. F. Gao, Y. Miyamoto, and D. Zhang, On physical and thermochemical properties of high-purity Ti_3SiC_2, *Materials Letters* **55**, 61 (2002).

[18]Songlan Yang, Zheng Ming Sun, and Hitoshi Hashimoto, Synthesis of Ti_3SiC_2 powder from 1Ti/(1-x)Si/2TiC powder mixtures, *Journal of Alloys and Compounds* **368**, 318 (2004).

[19]H.R. Orthner, R Tomasi, and W. J. Botta F., Reaction sintering of titanium carbide and titanium silicide prepared by high-energy milling, *Materials Science and Engineering* **A336**, 202 (2002).

[20]Shi Bo Li, Hong Xiang Zhai, Yang Zhou et al., Synthesis of Ti_3SiC_2 powders by mechanically activated sintering of elemental powders of Ti, Si and C, *Materials Science and Engineering* **A 407**, 315 (2005).

[21]Jing Feng Li, Toshiro Matsuki, and Ryuzo Watanabe, Mechanical-Alloying-Assisted Synthesis of Ti3SiC2 Powder, *J.Am.Ceram.Soc.* **85** (4), 1004 (2002).

[22]Zheng Ming Sun, Songlan Yang, and Hitoshi Hashimoto, Ti_3SiC_2 powder synthesis, *Ceramics International* **30**, 1873 (2004).

[23]Eldon Poulsen, Safety-Related Problems in the Titanium Industry in the Last 50 Years, *JOM: The Member Journal of TMS* **50** (5), 13 (2000).

[24]R. Radhakrishnan, S. B. Bhaduri, and Jr. C. H. Henager, Analysis on the sequence of formation of Ti_3SiC_2 and Ti_3SiC_2/SiC composites, *Powder Metallurgy Technology Conference;* (1995).

[25]Shi-Bo Li, Jian-Xin Xie, Li-Tong Zhang et al., In situ synthesis of Ti3SiC2/SiC composite by displacement reaction of Si and TiC *Materials Science and Engineering* **A 381**, 51 (2004).

[26]Ida Kero, Luleå University of Technology, 2007.

[27]B. D. Cullity, 3 ed. (Addison-Wesley, Reading, MA, 1967).

[28]M. J. Dickson, *Journal of Applied Crystallography* **2** (176) (1969).

[29]Sowmya Arunajatesan and Altaf H. Carim, Synthesis of Titanium Silicon Carbide, *J.Am.Ceram.Soc.* **78** (3), 667 (1995).

[30]Roman Pampuch, Jerzy Lis, Ludoslaw Stobierski et al., Solid Combustion Synthesis of Ti_3SiC_2, *Journal of the European Ceramic Society* **5**, 283 (1989).

[31]R. Radhakrishnan, J. J. Williams, and M. Akinc, Synthesis and high-temperature stability of Ti_3SiC_2, *Journal of Alloys and Compounds* **285**, 85 (1999).

[32]Jerzy Lis, Y. Miyamoto, Roman Pampuch et al., Ti3SiC (sic!) - based materials prepared by HIP-SHS techniques, *Materials Letters* **22**, 163 (1995).

[33]José M. Córdoba, María J. Sayagués, María D. Alcalá et al., Synthesis of Ti_3SiC_2 Powders: Reaction Mechanism, *J. Am. Ceram. Soc.* **90** (3), 825 (2007).

[34]Ida Kero, Marta-Lena Antti, and Magnus Odén, Time and temperature dependence of Ti_3SiC_2 formation from TiC/Si powder mixtures, *Submitted to Powder Technology* (2008).

[35]J. L. Ratliff and G. W. Powell, *Research on Diffusion in Multiphase Ternary Systems.* (National Technical Information Service, Alexandria, VA, 1970).

TOUGHENING OF A ZrC PARTICLE-REINFORCED Ti$_3$AlC$_2$ COMPOSITE

G. M. Song[a,*], Q. Xu[b], W. G. Sloof[c], S. B. Li[d], S. van der Zwaag[a]

[a] Faculty of Aerospace Engineering, Fundamentals of Advanced Materials, Delft University of Technology, Kluyverweg 1, 2629HS Delft, The Netherlands
* G.Song@tudelft.nl, Tel: 0031-15-2789459, fax: 0031-15-2786730
[b] National Center for High Resolution Electron Microscopy, Delft University of Technology, Lorentzweg 1, 2628 CJ Delft, the Netherlands
[c] Department of Materials Science and Engineering, Delft University of Technology, Mekelweg 2, 2628CD Delft, The Netherlands
[d] Materials Engineering Center, Beijing Jiaotong University, Beijing 100044, P. R. China

ABSTRACT

The fracture toughness of a ZrC particle-reinforced Ti$_3$AlC$_2$ ceramic composite (ZrC/Ti$_3$AlC$_2$) was measured to be as high as 11.5 MPa·m$^{1/2}$, which is 47% higher than that of the monolithic Ti$_3$AlC$_2$. A microstructural analysis with SEM and TEM showed that during hot pressing the ZrC was partially decomposed leading to Zr being dissolving into the Ti$_3$AlC$_2$ matrix to form a Ti$_x$AlC$_2$ solid solution containing Zr. On the other hand Ti and Al were dissolved into the ZrC to form a (Al,Ti,Zr)C solid solution. Observations on the cracked composite showed that crack deflection, crack bridging, grain pullout operating in the matrix mainly account for the high toughness of the material. The crack paths became more serious zigzag when the crack met ZrC particles. Residual thermal stress played a key role on the toughness increment of the composite. The solid solution induced toughening might also account for the high toughness of the composite. The unique combination of high strength, high toughness and good machineability makes the novel ZrC/Ti$_3$AlC$_2$ composite very attractive for high temperature applications such as high-precision electro-thermo-mechanical components.

INTRODUCTION

The currently developed layered Ti$_3$AlC$_2$ ternary carbide has a good high temperature strength, high elastic modulus, good oxidation resistance, high thermal shock resistance, low density, good electrical/thermal conductivity, and is readily machinable with conventional tools [1 - 3].The combination of all these good properties makes Ti$_3$AlC$_2$ an attractive candidate for high temperature applications. However, the inherent brittleness and relatively low strength of Ti$_3$AlC$_2$ may hinder its potential application in structural components. The highest reported fracture toughness of monolithic Ti$_3$AlC$_2$ prepared by Barrsum et al [1] or Wang et al [2] measured with single notch beam bending method was 7.2 MPa·m$^{1/2}$. Improvements in strength and toughness of Ti$_3$AlC$_2$ are therefore desirable to enable the application of Ti$_3$AlC$_2$ in practical applications, especially when these improvements can be achieved without sacrificing the other nice properties of Ti$_3$AlC$_2$ such as the thermal and electrical conductivity and machinability, etc. One possible way for improving the toughness of Ti$_3$AlC$_2$ is by introducing discrete reinforcements, such as Al$_2$O$_3$ [4,5], TiB$_2$ [6], SiC [7], ZrO$_2$ [7] and Cu particles [8,9]. Al$_2$O$_3$ and ZrO$_2$ are isolators and consequently will decrease the electrical conductivity of Ti$_3$AlC$_2$. Cu is electroconductive, but it cannot work well at high temperatures due to its low melting point (1184.62 °C) and its low hardness. The addition of 10% of TiB$_2$ resulted in a drop in fracture toughness to 6.6 MPa·m$^{1/2}$ [6], while the fracture toughness of Ti$_3$AlC$_2$ containing 20 wt% SiC was measured to be still 7.2 MPa·m$^{1/2}$[7].

In the present paper we explore a potential of toughening Ti_3AlC_2 by the addition of ZrC. ZrC was selected as a potentially attractive toughening agent as ZrC is an electric conductor (electrical resistance: 0.43 $\mu\Omega\cdot$m [10], which is comparable with that of Ti_3AlC_2, 0. 34$\mu\Omega\cdot$m [1]), and ZrC is expected to have a good chemical compatibility with Ti_3AlC_2, since the dimensions of the cubic ZrC crystal lattice are very close to those of the octahedral Ti_3C_2 unit in Ti_3AlC_2 lattice [2,10]. No earlier work on the reinforcement of Ti_3AlC_2 with ZrC has been reported. The ZrC particle-reinforced Ti_3AlC_2 (ZrC/Ti_3AlC_2) composite was prepared via hot-pressing method, and the microstructure, cracking behavior and the relevant toughening mechanisms of the new ZrC/Ti_3AlC_2 composite were investigated.

EXPERIMENTAL

The ZrC/Ti_3AlC_2 composite was fabricated via a solid-liquid reactive hot-pressing method. Ti (~48 μm, >99% purity), Al (~75 μm, >99% purity) and graphite powders (~45 μm, >99% purity) with a molar ratio of 3:1.1:2 for the Ti_3AlC_2 matrix, and ZrC powder (~3 μm, >98% purity) as the reinforcement were mixed by ball milling for 20 hrs in an ethanol solution. The volume ratio of Ti_3AlC_2:ZrC is 0.8:0.2. The slurry was dried at 60 °C and cold-pressed into lumps in graphite dies lined with BN layers under 8 MPa, and then hot-pressed at 1500 °C under 30 MPa for 2 hrs in an argon atmosphere. During hot pressing, a solid-liquid reaction occurred among solid Ti, C and liquid Al to produce a Ti_3AlC_2 matrix with ZrC particles homogenously distributed. For comparison, a monolithic Ti_3AlC_2 sample was also prepared.

Three point bending method was used to test the fracture toughness and fracture strength. The single-edge notched beam (SENB) is convenient for the measurement of fracture toughness although this technique shows a tendency to over estimate fracture toughness in some cases. SENB test bars, having dimensions of 2×4×20 mm³, were cut from the sintered ZrC/Ti_3AlC_2 composite lumps with a diamond blade and then polished with SiC sand paper followed by a diamond paste polish. A half-thickness notch with 2 mm in depth for each test bar was made at one side using a diamond saw with a thickness of 100 μm. The fracture toughness test to determine K_{IC} was performed with a 16 mm span at 0.05mm/min crosshead speed on a universal testing machine (Zwick/Roel 1455, Germany). The maximum load upon fracture P was used to determine the K_{IC} value using the well known equation [11],

$$K_{IC} = \frac{3PS}{2BW^2}\sqrt{a}\left[1.93 - 3.07\frac{a}{W} + 14.53\left(\frac{a}{W}\right)^2 - 25.11\left(\frac{a}{W}\right)^3 + 25.80\left(\frac{a}{W}\right)^4\right], \qquad (1)$$

where a is the crack length, B is the specimen width, W is the specimen thickness, S is the span length. An average value of K_{IC} was obtained from the tests run on at least six specimens. The critical strain energy release rate, G_{IC} was calculated from the stress intensity value using the relationship:

$$G_{IC} = \frac{(1-v^2)K_{IC}^2}{E}, \qquad (2)$$

where v is Poisson's ratio and E is Young's modulus. In some cases it was possible to unload the sample at a point just after the load drop following the maximum load because the sample did not split into two unconnected parts. In these cases a well defined critically developed crack process zone at the crack tip was obtained enabling detailed microstructural investigations.

Specimens with dimensions of 1.5mm×2 mm×20mm were used for flexural strength (σ_f)

measurement by the three-point bending method, with a 16 mm span at a 0.5mm/min crosshead speed. Vickers hardness (HV) was measured with a Vickers hardness tester (Buehler Omnimet MHT 7.0, USA) under a constant load of 9.8 N for 15 s.

In order to reveal the toughening mechanisms, the fracture surfaces as well as the critical crack zone at crack tips of the tested SENB samples were characterized with scanning electron microscopy (SEM, JSM 6500F, Japan). The microstructures were characterized on a transmission microscope (FEI Tecnai200 S-TWIN, the Netherlands) having a field emission gun and operated at 200kV. Chemical composition analyses were conducted with energy dispersive spectroscopy (EDS) under STEM mode with spot size less than 0.5 nm. Sample drift was checked before and after acquiring all spectra. The phase identification was conducted using X-ray diffraction with Cu Ka radiation (XRD, Bruker AXS D5005, Germany).

RESULTS AND DISCUSSION

Microstructure

SEM observation on the polished surface of the composite showed that the distribution of ZrC particles in Ti₃AlC₂ matrix is uniform except for a small amount of agglomerates of finer ZrC particles; see Fig. 1. Some Al_2O_3 particles were also present. Small amount of Al_2O_3 impurities were not only detected in the composite, but also in the monolithic Ti₃AlC₂, suggesting that the Al_2O_3 impurity resulted from the oxidation of Al powders during the preparation process of the composite. Previous research [1,4] showed that the presence of small amounts of Al_2O_3 is beneficial to the strength and toughness of Ti₃AlC₂.

Fig. 2 is a typical TEM image of the ZrC/Ti₃AlC₂ composite, showing the morphologies of ZrC particles within the Ti₃AlC₂ matrix. The chemical composition analysis with EDS along a line more or less perpendicular to the ZrC/Ti₃AlC₂ interface showed that interdiffusion had occurred at the interface zone. Some Zr was dissolved into the Ti₃AlC₂ matrix, while some Ti and small amounts of Al were dissolved within the surface zone of ZrC particles to form (Zr.Ti.Al)C solid solution. The Zr content in Ti₃AlC₂ matrix were measured with EDS as high as 22 at% while Al and Ti dropped to 11 at% and 64 at% respectively. It should be noted that the Zr content in the matrix is higher than the Al content. The molar ratio of Ti:(Zr+Al)=1.94, and (Ti+Zr):Al=7.64, seeming that Zr dissolved in the Ti₃AlC₂ matrix by partially substituting Ti and also partially substituting Al so as to keep a M_3AX_2 formula, perhaps forming a $(Ti,Zr)_3(Al,Zr)C_2$ solid solution. Further precise measurement of the composition of Zr in Ti₃AlC₂ is needed because EDS measurement is of qualitative. The interface interdiffusion should enhance the interface adhesion. The XRD analysis of the composite showed that the main phases in the composite are indeed ZrC and Ti₃AlC₂, which is consistent with the SEM and TEM observations.

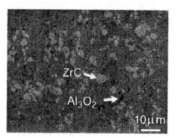

Fig. 1. Back scatting image of the ZrC/Ti₃AlC₂ composite, showing an uniform distribution of ZrC particles (gray phase) in Ti₃AlC₂ matrix containing small amount of Al_2O_3 particles (dark phase).

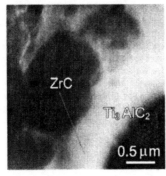
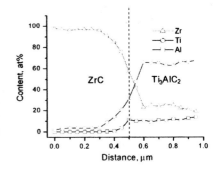

Fig. 2. TEM bright-field image of ZrC/Ti₃AlC₂ composite and the relative content of Al, Ti and Zr along the line across the interface.

Mechanical Properties

The mechanical properties of the ZrC/Ti₃AlC₂ composite are listed in Table 1. The average fracture toughness of the composite is as high as 11.5 MPa·m$^{1/2}$, a 47% higher than that of the monolithic Ti₃AlC₂, which means that the toughening effect of ZrC particles is significant. The critical strain energy release rate, G_{IC}, of the composite is 412 J/m^2, which is double of that of the monolithic Ti₃AlC₂, implying that more energy is needed for the crack propagation in the composite. The flexural strength of the composite are 490 ± 30 GPa, a 32% increase in flexural strength compared with the monolithic Ti₃AlC₂. Meanwhile, the elastic modulus of the composite is also slightly higher than that of monolithic Ti₃AlC₂, which can be attributed to the higher elastic modulus of ZrC (~400 GPa [10]).

Table 1 Properties of monolithic Ti₃AlC₂ and ZrC/Ti₃AlC₂ composite

Properties	Ti₃AlC₂	ZrC/Ti₃AlC₂
Elastic modulus (E), GPa	296 ± 15	311 ± 48
Flexural strength (σ), MPa	370 ± 20	490 ± 30
Fracture toughness (K_{Ic}), MPa·m$^{1/2}$	7.8 ± 0.4	11.5 ± 1.0
Critical strain energy release rate (G_{Ic}), J/m^2 *	198 ± 34	412 ± 54

* $v = 0.2$ is used for the calculation of G_{IC} [12].

Cracking Behavior

Fig. 3 shows a load-deflection curve recorded on a SENB sample of the ZrC/Ti₃AlC₂ composite during three-point bending. The present curve, showing a sawtooth feature after the maximum load, differs significantly from the linear-elastic behavior showing no residual strength after beyond the maximum load, demonstrated by normal brittle materials. The SEM observation of the cracks in the failed samples revealed extensive bridging and pullout of lamellar Ti₃AlC₂ grains on the crack propagation path, as shown in Fig. 4a. The existence of the crack bridging zone behind the crack tip decreases the stress intensity at the crack tip. A further propagation of the crack then requires a higher load. The extension of a crack therefore proceeds in an irregular manner by repeated initiation and arrest, and the load-displacement curve exhibits a sawtooth shape (Fig. 3). The similar nonlinear failure features of monolithic layered Ti₃AlC₂ and Ti₃SiC₂ have been reported elsewhere [1,2,13-15], which means the nonlinear feature is due to the Ti₃AlC₂ matrix of the composite. The crack bridging by ZrC particles or crack pinning occurred sporadically. Any way, it is reasonably to believe that the contribution from the bridging and pullout of lamellar Ti₃AlC₂ grains is much higher than that from the crack bridging and crack pinning caused by ZrC particles.

The zigzag crack growth mode resulted in a remarkable crack deflection (Fig. 4). The crack deflection became more severe as ZrC was present in the crack path. When impinging the ZrC particles, the crack preferred to cross through the particles and then propagate in the matrix or deflect an angle, rather than run at the interface between the matrix and the particles (Fig. 4b), and thus the crack path became more tortuous and, according to the crack deflection model [16,17], the energy consumption increases. This zigzag crack pattern in combination with local crack bridging and grain pullout is responsible for the high fracture toughness of the matrix and the composite.

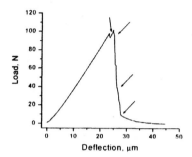

Fig. 3. Load-deflection curve of ZrC/Ti₃AlC₂ composite. showing a nonlinear feature. single-edge notch beam

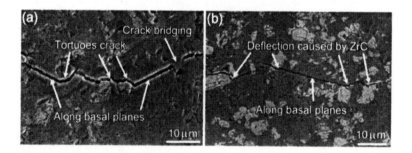

Fig. 4. Crack path in the ZrC/Ti₃AlC₂ composite. Crack preferred to propagate along the basal planes of the matrix, and meanwhile was deflected when it crossed through ZrC particles

Fracture Morphologies

Typical fracture surfaces of the ZrC/Ti₃AlC₂ composite are shown in Fig. 5 and 6. There are two types fracture at the fracture surface of the matrix: intergranular and transgranular fracture, the latter being predominant; see Fig. 5a. The transgranular fracture consisted of the delamination along the basal planes of the hexagonal Ti₃AlC₂ grains (indicated by arrows in Fig. 5a) and the breaking of the delaminated lamellas (indicated by arrows in Fig. 5b), which is consistent with the observation on the crack path (see Fig. 4). Due to the weak bonding between the Al atom layer and octahedral Ti₃C₂ layer governs by the metallic bond of Ti-Al, the crack will be deflected repeatedly on the inner basal planes in the lamellar Ti₃AlC₂ grains, and the crack path becomes severely zigzagged also on a submicron scale.

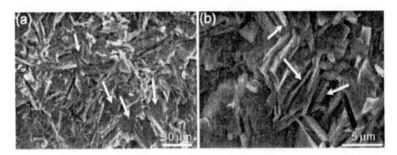

Fig. 5. Fracture surface of the ZrC/Ti₃AlC₂ composite, showing a predominant transgranular fracture. The arrows indicate the delamination along the Ti₃AlC₂ basal planes in (a) and the breaking of delaminated Ti₃AlC₂ lamellas in (b).

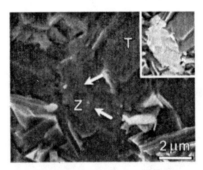

Fig. 6. Fracture surface of the ZrC/Ti₃AlC₂ composite. The interface of ZrC to Ti₃AlC₂ matrix is identified with backscattering image inserted. The river patterns, ridges and prongs of ZrC particle are indicated with arrows. The letters "T" and "Z" represent T₃AlC₂ and ZrC, respectively.

The fracture surfaces of the ZrC particles exhibited features of cleavage: transgranular fracture facets with occasional fragments of river patterns, and small amounts of thin break-off ridges and prongs, as indicated by the arrows in Fig. 6. The cleavage of ZrC particles implies that when the propagating crack meets a ZrC particle, the crack most frequently crossed through the ZrC rather than propagated around the particle surface, and this is consistent with the observation on the crack path in Fig. 4. A conclusion may be drawn as that the interface between the particle and the matrix is strong. Such an interface ensures the stress to be effectively transferred from the Ti_3AlC_2 matrix to the harder ZrC particles.

Toughening mechanisms

In particle-reinforced ceramics, residual stress, crack deflection and crack bridging are seen as the main toughening mechanisms [18]. For this ZrC/Ti_3AlC_2 composite, the toughening mechanisms at least consists of crack bridging, grain pullout and crack deflection, and these mechanisms are imposed by the lamellar Ti_3AlC_2 grains and are present in the monolithic and composite material alike. However, the composite exhibits a remarkably higher toughness than the matrix, which implies other toughening mechanisms operate in the composite beside the three mechanisms indicated above.

Beneficial residual stresses may have been induced by the thermal expansion misfit between the ZrC particles and Ti_3AlC_2 matrix when the composite was cooled down from its fabrication temperature because the expansion coefficient is 6.7×10^{-6}/K for ZrC particle [10], and 9.0×10^{-6}/K for Ti_3AlC_2 matrix [19], respectively. A radial compressive stress at the ZrC/Ti_3AlC_2 interface and a tangential tensile stress in the surrounding Ti_3AlC_2 matrix arise, if the composite is considered as a single sphere particle imbedded in an infinite elastic matrix [20]. The compressive stress at ZrC/Ti_3AlC_2 interface enhances the interface bonding, which ensures the hard ZrC particles to withstand higher stress loading by effectively transferring load from the matrix to the particles at the interface, and consequently, an effective strengthening effect of ZrC particle on Ti_3AlC_2 matrix is achieved. A residual tensile stress in the matrix might favor crack branching and crack deflection and micro-cracking in the matrix. So the residual stress favors the toughening effect of ZrC particles. The uniform residual stress σ within the particle can be approximately calculated as [20]

$$\sigma = \frac{(\alpha_m - \alpha_p) \cdot \Delta T}{\left(\dfrac{1 + v_m}{2E_m} + \dfrac{1 - 2v_p}{E_p} \right)}, \tag{3}$$

where a, E and v represent the thermal expansion coefficient, Young's modulus and Possion's ratio, respectively. ΔT is the temperature change between the processing and room temperature. The subscripts m and p stand for matrix and particle, respectively. The brittle-ductile transition temperature of Ti_3AlC_2 matrix is 1050 °C [19], so ΔT could be about 1025 °C. Furthermore, using $v_m = 0.2$ [1], $v_p = 0.19$ [10], $E_m = 297$ GPa, and $E_p = 400$ GPa [10], we obtain compressive residual stress, $\sigma = 660$ MPa, within the particle. The toughening effect caused by the residual stress in particle-reinforced composite is simply evaluated with the approach of Taya et al [21]:

$$\Delta K_{res} = 2\sigma \sqrt{2(\lambda - d)/\pi}, \tag{4}$$

in which λ is the center to center nearest neighbor spacing between particles, and can be approximately given as $\lambda = 1.083d / \sqrt{f_p}$, where f_p is the volume fraction of the particle. d is the average diameter of the particles, is taken as 1.5 μm. The resulting residual stress induced toughening effect is then estimated to be $\Delta K_{res} = 1.5$ MPa·m$^{1/2}$.

The existence of ZrC also resulted in more crack deflection in the composite. The crack surface area is enlarged by crack deflection, so the toughening contribution associated with the crack surface area can be calculated as follows. Assuming the maximum deflection distance, l_d, the resultant toughness from the addition surface area can be given by assuming that the fracture morphology comprises cones (height, 0-l_d) [16],

$$\frac{G_d}{G_m} = \frac{2}{\pi}\frac{2}{\lambda}\int_0^{\pi/2}\sqrt{\left(\frac{\lambda}{2}\right)^2 + (l_d \sin\theta)^2}\,d\theta \tag{5}$$

where G_d and G_m are the strain energy release rates of the deflected and undeflected crack, respectively. The crack deflection contribution, ΔK_d, is simply given by,

$$\Delta K_d = \left(\sqrt{G_d/G_m} - 1\right)K_m. \tag{6}$$

ZrC induced crack deflection not only occurred within the matrix but also occurred within the ZrC particles. The average deflection distance, l_d, is less than or equals the radius of the particle. Here, l_d is taken as 0.5 μm. Thus, ΔK_{res}=0.5 MPa·m$^{1/2}$. Assuming the final effective fracture toughness to be the linear addition of the three contributions,

$$K_{IC} = K_m + \Delta K_{res} + \Delta K_d \tag{7}$$

the calculated toughness of the composite is about 9.8 MPa·m$^{1/2}$, which is still lower than the measured value, 11.5 MPa·m$^{1/2}$. The difference is provisionally attributed to Zr solid solution induced toughening of the Ti₃AlC₂ matrix. A more precise set of experiments to demonstrate the occurrence of solid solution toughening in this system will be conducted in the near future.

CONCLUSIONS

(1) The hardness, flexural strength and fracture toughness of the ZrC particle-reinforced Ti₃AlC₂ composites are measured as 7.8 GPa, 490 MPa and 11.5 MPa·m$^{1/2}$, respectively, demonstrating a significant strengthening and toughening effect due to the presence of ZrC particles.
(2) The microstructural observation on cracked samples shows that three toughening mechanisms, crack deflection, crack bridging and pullout caused by lamellas Ti₃AlC₂ grains operate both in the monolithic material and in the composite, accounting for the high toughness of the matrix and the composite.
(3) ZrC was partially dissolved into the Ti₃AlC₂ matrix to form Ti₃AlC₂ solid solution containing Zr, and Ti and Al were dissolved in ZrC to form (Al,Ti,Zr)C solid solution, which enhanced the interface adhesion.
(4) The toughness increment induced by the residual thermal stress in the composite was calculated at 1.5 MPa·m$^{1/2}$. The toughness contribution from the crack deflection caused by ZrC was estimated as 0.5 MPa·m$^{1/2}$. Solid solution toughening is held responsible for the unaccounted toughness increment of the composite to the monolithic Ti₃AlC₂.

ACKNOWLEDGEMENTS

This research was carried out as part of the Delft Center for Materials Research Program on Self Healing Materials.

REFERENCES

[1] N. V. Tzenov and M. W. Barsoum, Synthesis and characterization of Ti_3AlC_2. *J. Am. Ceram. Soc.*, **83**[4], 825-832 (2000).

[2] X. H. Wang and Y. C. Zhou, Microstructure and properties of Ti_3AlC_2 prepared by the solid-liquid reaction synthesis and simultaneous in-situ hot pressing process. *Acta Mater.*, **50**[12], 3141-3149 (2002).

[3] G. M. Song, Y. T. Pei, W. G. Sloof, S. B. Li and J. Th. M. De Hosson, S. van der Zwaag, Oxidation induced crack healing of Ti_3AlC_2 ceramics. *Scripta Mater.*, **58**[1], 13-16 (2008).

[4] J. X. Chen and Y. C. Zhou, Strengthening of Ti_3AlC_2 by incorporation of Al_2O_3. *Scripta Mater.*, **50**[6], 897-901 (2004).

[5] J. X. Zhou, Y. C. Zhou, H. B. Zhang, D. T. Wan and M. Y. Liu, Thermal stability of Ti_3AlC_2/Al_2O_3 composites in high vacuum. *Mater. Chem. Phys.*, **104** [1], 109-112 (2007).

[6] X. Lu, B. Mei, W. Zhou and C. Tian, Preparation and properties of Ti_3AlC_2/TiB_2 composites by spark plasma sintering. *J. Chinese Ceram. Soc.*, **35**[5], 648-652 (2007).

[7] L. M. Peng, Preparation and properties of ternary Ti_3AlC_2 and its composites from Ti-Al-C powder mixtures with ceramic particulates, *J. Am. Ceram. Soc.*, **90**[4], 1312-1314 (2007).

[8] J. Zhang, J. Y. Wang, Y. C. Zhou, Structure stability of Ti_3AlC_2 in Cu and microstructure evolution of Cu-Ti_3AlC_2 composites. *Acta Mater.*, **55**[13], 4381-4390 (2007).

[9] H. X. Zhai, M. X. Ai, Z. Y. Huang, Y. Zhou, S. B. Li and Z. L. Zhang, Unusual microstructures and strength characteristics of Cu/Ti_3AlC_2 cermets. *Key Eng. Mater.*, **336-338**, 1394-1396 (2007).

[10] H. O. Pierson, Handbook of refractory carbides and nitrides, Noyes Publications, USA, 1996.

[11] H. Tada, P. C. Paris, G. R. Irwin, The stress analysis of cracks handbook. Del Research Corporation, Hellertown, PA. 1973.

[12] Y. C. Zhou, Z. M. Sun, X. H. Wang, S. B. Chen, Ab initio geometry optimization and ground state properties of layered ternary carbides Ti_3MC_2 (M = Al, Si and Ge). *J. Phys. Con. Matter.*, **13**[44], 10001-10010 (2001).

[13] Y. W. Bao, J. X. Chen, X. H. Wang and Y. C. Zhou, Shear strength and shear failure of layered machinable Ti_3AlC_2 ceramics, *J. Eur. Ceram. Soc.*, **24** (5), 855-860 (2004).

[14] B. J. Kooi, R. J. Poppen, N. J. M. Carvalho, J. Th. M. De Hosson, M. W. Barsoum, Ti_3SiC_2: A damage tolerant ceramic studied with nanoindentations and transmission electron microscopy. *Acta Mater.*, **51**, 2859-72 (2003).

[15] Y. C. Zhou, Z. M. Sun, Microstructure and mechanism of damage tolerance for Ti_3SiC_2 bulk ceramics, *Mat. Res. Innovat.*, **2**, 360-363 (1999).

[16] K. T. Faber, A. G. Evans, Crack deflection processes-1. Theory. *Acta Metall.* **31** (1983) 565-576.

[17] K. T. Faber, A. G. Evans, Crack deflection processes-2. Experiment. *Acta Metall.* **31** (1983) 577-584.

[18] Y. Zhou, Ceramic Materials, Harbin Institute of Technology Press, China, 1995.

[19] H.B. Zhang, Y.C. Zhou, Y.W. Bao, and M.S. Li, Abnormal thermal shock behavior of Ti_3SiC_2 and Ti_3AlC_2, *J. Mater. Res.*, **21**, 2401-07 (2006).

[20] J. Selsing, Internal stresses in ceramics. *J. Am. Ceram. Soc.* **44**, 419 (1961).

[21] M. Taya, S. Hayashi, A. S. Kobayashi, H. S. Yoon, Toughening of a particulate-reinforced ceramic-matrix composite by thermal residual stress. *J. Am. Ceram. Soc.* **73**, 1382-91 (1990).

MICROSTRUCTURE AND PROPERTIES OF THE CERMETS BASED ON Ti(C,N)

S. Q. Zhou[1,2]*, W. Zhao[1], W. H. Xiong[1]
1 ⊏ Huazhong University of Science and Technology
Wuhan Hubei 430074, China
2 ⊏ The University of Waterloo
Waterloo ON, N2L3G1, Canada

ABSTRACT

Effect of Mo and Mo_2C on the microstructure and properties of Ti(C,N)-based cermets was investigated in this paper. The results have indicated that the weight percentage of Mo from 5 to 10 can reduce Ti(C,N) grain diameter and thickness of the rim, and Ti(C,N) grain can be wetted by Ni-Cu-Mo liquid so as to getting small contiguity of Ti(C,N) grain. In that way, the transverse rupture strength of Ti(C,N)-based cermets has reached to 1800-1900MPa, the fracture toughness has been due to 16-18MPa $m^{1/2}$. But 15 wt% Mo was not more effective on Ti(C,N)-based cermets, because the thickness of the rim becomes larger. In the circumstance of Mo_2C, 5 wt% Mo_2C was good for microstructure and properties of Ti(C,N)-based cermets, but 11 wt% Mo_2C has resulted in larger contiguity of Ti(C,N) grain and big Ti(C,N) grain diameter so as to reducing transverse rupture strength and fracture toughness. So that, the effect of Mo on Ti(C,N)-based cermets is better than Mo_2C.

INTRODUCTION

Titanium carbonitride-based (Ti(C,N)-based) cermets are important structural and wear-resistant materials which are widely appreciated in metal cutting application, owing to their excellent mechanical properties [1-3]. Recently, more and more conventional WC–Co-based hard alloys are being replaced by Ti(C,N)-based cermets, accompanied with the trend of high speed machining. Comparing with WC–Co hard metal, the advantages of Ti(C,N)-based cermets lie in their higher hot hardness, wear resistance, chemical stability, and resistance to plastic deformation at elevated temperature.
Consequently, cermets cutting tools show improved surface finishing and tolerance control. Meanwhile, the cutting efficiency and tool life are improved too. On the whole, in high speed finishing and semi-finishing cutting applications, Ti(C,N)-based cermets are more preferred than WC–Co hard metals [4-6]. The complex microstructure of Ti(C,N)-based cermets is the key to interpret their mechanical properties, so it has been widely studied by many authors for years [7-9]. Generally, cermets material is composed of two phases: one is the ceramic phase (titanium carbonitride hard phase) and the other is metal binder phase (nickel or cobalt or a mixture of them) which bonds the ceramic phase. In general, ceramic phase provides high hardness to this class of materials, while the metal binder phase contributes to ductility, toughness and thermal-shock resistance. Microstructure examination shows that a typical carbonitride grain often demonstrates a core/rim structure: black Ti(C,N) cores surrounded by grey (Ti, W, Mo, Nb, Ta, . . .) (C, N) complex carbonitride rims (in scanning electron microscopy (SEM)–back scattered electron (BSE) contrast), resulting from a dissolution– repricipitation process.

Previous work has shown that Mo or Mo_2C is effective chemical continent for improving wettibility [10-12] and by forming (Mo,Ti)(C,N ⊏ in the surface of Ti(C,N) particle, they may improve interface bonding between Ti(C,N) and Ni. But there is argument about their amount and initial powder statue, the excellent properties can be obtained while the content of Mo is 10wt%-20wt% [13]. Also it was reported that Mo_2C may reduce grains and increase properties of the cermets by forming (Mo,Ti)(C,N) rim phase in solid sintering, which prohibits Ti(C,N) dissolution in liquid sintering [14] and the excellent properties can be obtained while adding amount of Mo_2C is due to10wt% [15].

In the cermets _ ceramic particles would contact each other, the contact area percentage is defined as contiguity. In WC/Co hard alloys, the parameter contiguity is defined as[16]:

$$C=\frac{2S_{cc}}{2S_{cc}+S_{cm}} \tag{1}$$

Where S_{cc} is the area between the carbon particles and S_{cm} is the area between the carbide particles and the matrix in a unit volume.

ASTM provides the recommend practice for estimation of the area fraction of a phase in a multiphase alloy by systematic point counting on a planar test section. The point fraction, $(P_P)_a$ intercepted by the a phase, statistically interpreted, provides an estimate of the volume fraction, $(V_V)_a$, area fraction, $(A_A)_a$, and line fraction, $(L_L)_a$, For random measurements.

$$(P_P)_a=(L_L)_a=(A_A)_a=(V_V)_a \tag{2}$$

The other structure parameters are obtained from boundary intercepts with test lines on planar sections; one determines the average number of intercepts per unit length of test line with traces of the carbide/cobalt interface, $(N_L)_{WC/Co}$, and of carbide/ carbide grain boundaries, $(N_L)_{WC/WC}$. From these quantities, one can calculate the average carbide grain size:

$$D_{WC}=L_{WC}/(N_L)_{WC/WC} \tag{3}$$

the contiguity of the WC phase:

$$C=2(N_L)_{WC/WC}/[2(N_L)_{WC/WC}+(N_L)_{WC/Co}] \tag{4}$$

And the mean free path in the binder phase:

$$\lambda_M=L_{Co}/(N_L)_{WC/Co} \tag{5}$$

WC/Co composites have been studied extensively and some useful relationship among hardness and contiguity and the mean free path have been obtained:

$$H=H_{WC}CV_{WC}+H_{Co}(1-CV_{WC}) \tag{6}$$

where H is the hardness of the composites; H_{WC} and H_{Co} are the hardness of WC and Co, respectively; C is the contiguity of the WC particles, and V_{WC} is the volume fraction of WC in the composites.

The hardness of cobalt was related to the mean free path(λ) of the cobalt matrix in the composites by the following empirical relation:

$$H_{Co}=304+12.7\lambda^{-0.5} \tag{7}$$

The homogeneous distribution of a high volume fraction of fine WC particles in a cobalt matrix enhances strength, wear resistance, thermal conductivity and so on of the composites, but at the expense of strength and toughness [17]:

$$\sigma_Y=\sigma_b(1-CV_{WC})+\sigma_wCV_{WC} \tag{8}$$

where σ_b and σ_w are the in situ yield stress of Co and WC respectively. And :

$$C=1.03exp[-5(1-V_{WC})] \tag{9}$$

But the effect of alloy elements on contiguity can not be found in the above mentioned published reports. So that, in the present work, the effect of Mo and Mo_2C on mechanical properties and microstructures of Ti(C,N)-based cermets is investigated. Especially, the effect of Mo and Mo_2C on the contiguity of Ti(C,N)-based cermets and the relationship between contiguity and mechanical properties are analyzed based on experiments.

EXPERIMENTAL PROCEDURE
Specimen Preparation
    The characteristics of commercially obtained starting powders are listed in Table 1. After weighing and ultrasonic dispersion, raw powders were milled in a QM-1SP planetary ball mill. Blending was done at 180-200 rpm with WC–Co balls (ball-to-powder weight ratio, 7:1) for 24 h in ethanol bath. Then, the slurry mixture was dried for 12 h at the temperature of 80 ☐, after which dried powders were sieved

through No. 200 mesh and pelletized. Rectangular-shaped specimens were pressed at uniaxial pressure of 200 MPa for transverse rupture strength (TRS) and fracture toughness (K_{IC}) test. Finally, green specimens were dewaxed at 800 , and vacuum sintered (0.1 Pa) at 1450 for an hour. The sintering process curve is showed in figure 1. A series of five model systems were prepared by exactly the same method, and their nominal compositions are given in Table 2. The reason of selecting Cu as an alloy element was that Cu can dissolve into Ni completely, so that, melting temperature of powder press rough will fall, which is advantageous to reduce sintering temperature and thermal stress in cermets. A few WC and VC can diminish Ti(C,N) granule size and prevent the granule enlarging[18, 19]. So their content is controlled about 5wt%.

Table 1 The chemical composition of raw materials (wt%) and grain

material	C	N	S	O	W	Ti	Ni	Mo	Nb	d(μm)
TiC	18.800	—	0.0270	0.001720	—	Bal.	—	—	—	2.58
TiN	0.0890	24.9	0.0012	0.000135	—	Bal.	—	—	—	3.54
Ni	0.1780	—	0.0033	0.001060	—	—	Bal.	—	—	2.3
Mo	0.0094	—	0.0072	0.001215	—	—	—	Bal.	—	2.8
WC	6.11	—	0.0026	0.001020	Bal.	—	—	—	—	1.5

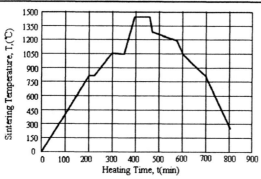

Fig.1 Sintering process projects in the experiment

Table 2 The contents of raw materials in experiment (wt%)

No.	TiC	TiN	Mo	Mo₂C	Cu	WC	VC	Ni
1	30	30	0	0	0	5	5	30
2	30	30	10	0	5	5	5	15
3	30	30	15	0	5	5	5	10
4	32	32	0	6	0	5	5	20
5	32	32	0	11	0	5	5	15

Experimental Methods

The microstructure of polished specimens (finished with 1 um diamond paste) was observed by FSEM (GENESIS, INC. USA,) and (JSE-6460, JEOL INC.) in BSE mode coupled with EDX (EDAX INC. USA,), and the fractured surface was observed in the secondary electron (SE) mode. Only metallic constituents of each sample were measured in this study since carbon and nitrogen content cannot be accurately quantified from SEM–EDX. Semi-quantitative information about grain size and phase volume fraction and the contiguity of ceramic particles was obtained by manual measurements on SEM micrographs using standard linear intercept and point counting methods(using equation 2-5), respectively. Phase identification of each system was carried out by XRD (X'Pert Data, Philips INC, USA). Three-point-bending TRS (20 mm span, 0.5 mm/min crosshead speed) was measured by using universal materials tester at room temperature and K_{IC} of the materials is calculated by the microscopic indentation method after the bending testing [20]. The specimen geometric sizes are 5 mm×5 mm×40 mm. Vickers indentation hardness with a load of 10 kg, 15 s loading time was measured.

EXPERIMENTAL RESULTS AND DISCUSSION
The Effect of Cu, Mo and Mo_2C on the Microstructures of Ti(C,N)-Based Cermets
    The phase constitutions of Ti(C,N)-based cermets before and after sintering are showed in figure 2. It can be found that the phase constitutions in sintered Ti(C,N)-based cermets is Ti(C,N) ceramic phase and Ni metal phase(as shown in Fig.2, 3), the result indicates that sintering procedure has been finished. But in case of Mo_2C, a few of Mo_2C ceramic phase also is available in the phase constitutions (as shown in Fig.2, 2). It can be proved that Mo_2C ceramic phase did not transform into (Ti, Mo) (C, N) Alloy carbonitride completely after 1450□ sintering 60 minutes.
    The effect of Cu and Mo & Mo_2C on grain diameter (d) and the contiguity(C) in Ti(C,N)-based cermets is showed in figure 3 and figure 4.
    According to figure 3 and 4, the grain diameter and the contiguity are reduced along with increasing the content of Cu and Mo, but in the case of Mo_2C, the results are different from Mo. When the amount of Mo_2C is more than 6wt%, the grain diameter and the contiguity increase. It can be presumed that Cu and Mo in the cermets can prevent granules from growing and also reduce the contiguity of ceramic grain. It can be found from figure 5 that the liquid phase temperature of system Ti(C,N)-Ni is reduced from 1320□ to 1270□ due to adding 5wt% Cu and 10wt%Mo. This improves the recomposition of ceramic particle ⌐formation of liquid phase and ceramic particles wetted by the liquid phase[21], so that, the effect of Mo restraining the grain growth and reducing the contiguity is more intense than that of Mo_2C.

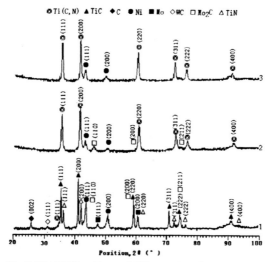

Fig. 2. The XRD patterns of different sintering status samples
1-Ni-Mo-Cu-TiC-TiN mixed powder(Before sintering); 2- Ni-Mo₂C-Cu-TiC-TiN(1450□ sintering 60 minutes); 3- Ni-Mo-Cu-TiC-TiN (1450□ sintering 60 minutes)

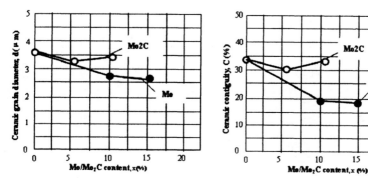

Fig.3. Effect of content of Mo and Mo₂C on grain diameter

Fig.4. Effect of content of Mo and Mo₂C on contiguity

The figure 6 a) is the microstructures of No 2 material in table 2 it is found that the particles recomposition has been finished and the grains are fine, and the contiguity of ceramic granules is small. On the contrary, when the content of Mo_2C is high, the ceramic granules will grow larger and the contiguity also increase as shown in figure 6 b)(Mo_2C added 11wt%). Because Mo_2C will diffuse into and react with Ti(C,N)(black core phase) so as to forming (Mo,Ti)(C,N)(grey rim phase) in solid sintering from 1000 -1250

which can accelerate ceramic granules contacting each other and increase solid sintering contiguity [22]. Simultaneously the dissolution of Ti(C,N) is stopped due to forming (Mo,Ti)(C,N) which results in the recomposition of ceramic particle and formation of liquid phase are more difficult, especially high content of Mo_2C. On the other hand, the more the carbide is, the higher the contiguity is [17,23], so the high content of Mo_2C results in high contiguity.

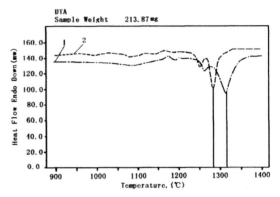

Fig.5. DSC testing results of part sample in table2
(1—No1 2—No2)

The Relationship between Microstructures and Properties of Ti(C,N)-Based Cermets
    The relationship among metal phase content, ceramic grain diameter & the contiguity and transverse rupture strength & fracture toughness is showed in figure 7 8 9 and 10.

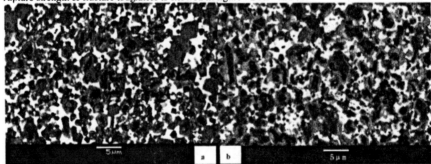

Fig.6 Effect of Mo and Mo₂C content on microstructure of Ti(C,N)-Ni system
a) Mo: 10wt%, b) Mo₂C: 11wt%

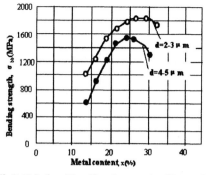

Fig.7. Relationship of bend strength with metal content

Fig.8. Relationship of fracture toughness with metal content

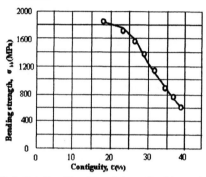

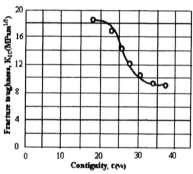

Fig.9. Relationship of bend strength with contiguity   Fig.10. Relationship of fracture toughness with contiguity

It can be found from figure 7 and 8 that transverse rupture strength & fracture toughness of Ti(C,N)-based cermets increase with metal phase content in the beginning, and then reach maximum at a suitable metal phase content(1820MPa at 27 wt% metal for 2-3µm granule diameter, 1580MPa at 24 wt% metal for 4-5 µm granule diameter) finally reduce as metal phase content increases. The larger the granule diameter is, the smaller the metals phase content of peak value TRS and fracture toughness is, transverse rupture strength is more sensitive than fracture toughness for the granule diameter.

The relationship between yielding strength and the volume percentage of ceramic phases & diameter of the ceramic granules can be described as equation 10[22]:

$$\sigma_y = (1000 + \frac{520 f_c^{2/3}}{\sqrt{d(1.26 - f_c^{1/3})}} + 1112 \frac{f_c^{1/3}(1 - 1.47 f_c)(2.25 f_c - 1)}{d(1.26 - f_c^{1/3})}) \sqrt{1 + 1.85 f_c} \qquad (10)$$

Where σ_y is yielding strength of Ti(C,N)-based cermets, f_c is the volume percentage of ceramic phases, d is diameter of the ceramic granules.

When d equals to 3 and 5 µm respectively, one can calculate the volume percentage of ceramic phases at maximum yielding strength using differential max method. The volume percentage of ceramic phases at maximum yielding strength is about 70 vol % for 3μm diameter and 75 vol % for 5μm diameter, respectively. The experiment results are in accord with the calculating values. The reason why yielding strength and fracture toughness appear peak value with metal phase content is that when metal phase content is high, the metal layer thickness between ceramic granules reaches a critical value which can build stress concentration in the metal phase, this stress leads that the crack germinates and expands in metals rapidly (also see fig.11), accordingly, which results in yielding strength and fracture toughness dropping.

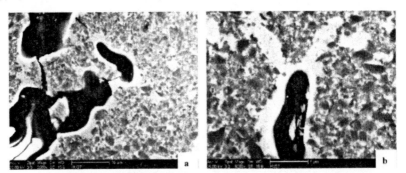

Fig.11. FSEM of crack developing path and crack tip in the content 33wt% of metal phase
a) Crack developing path, b) crack tip

The crack expanding and cracking tip shape in Ti(C,N)-based cermets which consists of 33 wt% metal phase are showed in figure 11. It can be observed that the crack propagates along with the path of thick metal phase. Although the crack stretching displacement is large, the cracking keeps away from ceramic granules (Fig.11a), this results in the nonavailability of the strengthening effect of ceramic granules and the passivated tip of crack proves that the toughening effect of metal phase is available(Fig. 11b). So that, the strength reduces distinctly, and fracture toughness decreases a little.

The relationship between contiguity and TRS & fracture toughness was showed in figure 9 and figure 10. It can be found that the smaller the contiguity is, the higher the transverse rupture strength & fracture toughness are. The transverse rupture strength was reduced with the contiguity as linear when contiguity is larger than 25%, but the fracture toughness breaks at 25% contiguity. When the contiguity is larger than 25%, the fracture toughness decreases rapidly and nears to 9 MPam$^{1/2}$ finally. While the contiguity is smaller than 25%, the fracture toughness increases rapidly and nears to 18.5 MPam$^{1/2}$ finally.

It can be observed by fracture surface of Ti(C,N)-based cermets that transgranular and intergranular fracture are most for the contiguity larger than 25%(as shown in Fig.12a). But when the contiguity is smaller than 25%, the tough hole and tearing fracture is majority (as shown in Fig.12b). The results indicate that the mode of fracture is changed near 25% contiguity, which means the mode of fracture is transferred from bridge toughening into crack shielding toughening [24].

It may be presumed that the contiguity is deleterious for transverse rupture strength & fracture toughness of Ti(C,N)-based cermets. But it will be reported in another paper that how the contiguity is controlled.

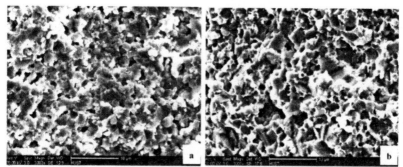

**Fig.12 FSEM of fracture face of Ti(C,N)-Ni composite on different contiguity
a) Contiguity: 32% , b)Contiguity:18%**

CONCLUSIONS

The effect of Mo and Mo_2C on microstructures and properties of Ti(C,N)-based cermets has been investigated in the paper. The results can be summarized as follows:

(1) The small granules and low contiguity of Ti(C,N) are produced with addition of Mo metal, meanwhile, the content of metal phase also increases.

(2) The liquid phase temperature of Ti(C,N) –cermets' powder compress is lowered with addition of Mo and Cu metal, which is advantageous to the liquid phase sintering.

(3) The optimizing mechanical properties are obtained at 10 wt% Mo, at this case, there are 25-28 wt% metal phase and 18-22% contiguity in the microstructures and the Ti(C,N) granule size is about 2-3 μm.

(4) The Mo_2C less than 5 wt% is advantageous to the microstructures and mechanical properties, but it increases Ti(C,N) granule diameter and contiguity so as to reducing mechanical properties when the addition amount of Mo_2C is larger than 5 wt%.

(5) The TRS has a peak value as a function of the metal phase at 27 wt % for 2-3μm granule diameter and at 24 wt% for 4-5 μm granule diameter, the peak value is 1820MPa and 1580MPa, respectively. The same as fracture toughness, the peak value appears at 25 wt% metal phase for 2-3μm granule diameter and at 20 wt% metal phase for 4-5μm granule diameter, the peak value is $18.4 MPam^{1/2}$ and $17.9 MPam^{1/2}$, respectively.

(6) The TRS is more sensitive than fracture toughness for the granule diameter.

(7) The TRS and fracture toughness are reduced greatly when contiguity of Ti(C,N) granules is larger than 25%. The reason that leads to the reducing is that brittle fracture occurs in the materials.

*Corresponding author. Tel.: +86 27 87541516 Fax: +86 27 87542504
E-mail address: zhousqwq@263.net

Supported by the National Natural Science Foundation of China (General Program No.50074017/E0408)

REFERENCES
[1]P. Ettmayer, H. Kolaska, W. Lengauer, and K. Dreyer, Ti(C,N) Cermets—metallurgy and properties. Int. J. Refr. Met. & Hard Mat. **13**,343 1995

[2] V. Richter, and M. Ruthendorf, Composition, microstructure, properties and cutting performances of cermets. In: Proceedings of Euro PM '99 Conference on Advances in Hard Materials Production. 229 1999 .

[3]Y. Zheng , M. You, W. H. Xiong, W. J. Liu , S. X. Wang, Valence-electron structure and properties of main phases in Ti(C, N)-based cermets. Materials Chemistry and Physics **82**, 877(2003).

[4] E. B. Clark and B. Roebuck, Extending the application areas for titanium carbonitride cermets. Refr. Met. & Hard Mat. **11,** 23 (1992)

[5] S. Zhang, Material development of titanium carbonitride-based cermets for machining application. Key Engineering Materials, **138–140,** 521 (1998).

[6] H.Pastor, Titanium-carbonitride-based hard alloy for cutting tools. Mat. Sci. Eng. **A105/106** 401(1988).

[7] D. S. Park and Y. D. Lee, Effect of carbides on the microstructure and properties of Ti(C,N) based ceramics. J. Am. Ceram. Soc. **82 (11),** 3150(1999).

[8] S. Y.Ahn, S. W. Kim, and S.Kang, Microstructure of Ti(C,N)–WC–NbC–Ni cermets. J. Am. Ceram. Soc. **84(4),** 843(2001).

[9] S. Y Ahn & S.Kang, Formation of core/rim structure in Ti(C,N)–WC–Ni cermets via a dissolution and precipitation process. J. Am. Ceram. Soc. **83 (6),** 1489(2000)

[10] W.Xiong, Phase and microstructure review of hard alloy mould materials. Materials Report **(5),** 24(1992)

[11] W. Xiong, F. Zhou, G. Li and K. Cui. The effect of powder granularity on microstructures and properties of cermets-based Ti(C,N).Transaction of Huazhong University of Science and Technology **23(12),** 37 (1995)

[12] Y. Hu. Preparation process and microstructure transformation of cermets-based Ti(C,N). PhD thesis. Library of Huazhong University of Science and Technology. 38 Wuhan, 2002 .].

[13] L. Liu, C. Liu, X. Zhao. The relationship between Mo Ni content and microstructures & properties of cermets-based Ti(C,N). Hard Alloy **11(2),** 74(1994)

[14] J. K. Park, S. T. Park, Densification of TiN-Ni cermets by improving wettability of liquid nickel on TiN grain surface with addition of Mo₂C. International Journal of Refractory Metals & Hard Materials. **17,** 295 (1999).

[15] D. Mari, S. Bolognini, G. Feusier, et al, TiMoCN based cermets Part I. Morphology and phase composition. International Journal of Refractory Metals & Hard Materials. **21,** 37(2003).

[16] K. Jia, T.E. Fischer and B. Gallois, Microstructure, Hardness and Toughness of Nanostructured and Conventional WC-Co Composites. Nanostructured Materials. **10(5),** 875(1998).

[17] Z. Fang. Correlation of transverse rupture strength of WC-Co with hardness. International Journal of Refractory Metals & Hard Materials. **23,** 119(2004).

[18] Y. Zheng, W. Liu, M. You. The effect of Cr₃C₂ and VC on value electronic structure and properties of rim phase in cermets –based Ti(C, N). Silicate Transaction. **32(4)** 422 (2004).

[19] W. T. Kwona, J. S. Park, S. W. Kimb,et al, Effect of WC and group IV carbides on the cutting performance of Ti (C,N) cermets tools ,International Journal of Machine Tools & Manufacture. **44,** 341(2004).

[20] D K Shetty. Indentation of WC/Co Cermets. J Materials Science, **20,** 1873 (1995).

[21] S. Y. Wang, W. H. Xiong, Microstructure and mechanical properties of Ti(C, N)-based cermets with different molybdenum contents. Trans. Nonferrous Met. Soc. China. **15(3),** 148. (2005)

[22] S. Q. Zhou, Investigation of interface and properties on ceramic matrix-metal composites. PhD thesis. Library of Huazhong University of Science and Technology, 33(Wuhan, 2006)

[23] M. A. Qian; L. C. Lim. On the disappearance of Mo2C during low-temperature sintering of Ti(C,N)-Mo₂C-Ni cermets. Journal of Materials Science. **34,** 3677 (1999)

[24] B. Budiansky, J. C. Amazigo and A. G. Evans, Small-scale crack bridging and the fracture toughness of particulate -reinforced ceramics. J. Mech. Phys. Solids, **36(2),** 167 (1988)

SCRATCH-INDUCED DEFORMATION AND RESIDUAL STRESS IN A ZIRCONIUM DIBORIDE-SILICON CARBIDE COMPOSITE

Dipankar Ghosh[a], Ghatu Subhash[a†] and Nina Orlovskaya[b]

[a]Department of Mechanical and Aerospace Engineering,
University of Florida, Gainesville, FL 32611, USA.

[b]Department of Mechanical, Materials and Aerospace Engineering,
University of Central Florida, Orlando, FL 32816, USA.

ABSTRACT

A ZrB_2-5wt%SiC composite was produced employing a plasma pressure compaction technique at a temperature of 1750°C and at a pressure of 43 MPa over a short consolidation period of only 5 min. The resulting compact had a sintered density above 96% of theoretical density. Scratch experiments were conducted on the composite in the load range of 50 to 250 mN using a Berkovich nanoindenter. Microstructural analysis of the scratch grooves revealed that the grooves consisted of several sets of slip lines oriented at random angles to the scratch direction and transgranular microcracks, oriented orthogonal to the scratch direction. Also, ZrB_2 grain-boundary fracture and interfacial cracking between ZrB_2 matrix and SiC particles were observed at higher scratch loads. An analytical model based on combined Boussinesq field and Cerruti field solutions was used to rationalize the formation of slip lines and orthogonal microcracks. It was postulated from the mechanistic analysis that the maximum shear stress which occurs ahead of the indenter tip during the scratch process caused the slip lines where as the principal tensile stress in the wake of the indenter caused the microcracking perpendicular to the scratch direction. This postulation was later verified by the observation of slip lines ahead of the scratch groove at the exit-end. Micro-Raman spectroscopy was employed to measure the scratch-induced residual stress within the SiC particles. The Raman spectra indicated a gradual increase in the residual stress with scratch load.

1. INTRODUCTION

Development of light weight and high strength materials which can withstand elevated temperatures (above 2000°C) and provide good thermal insulation properties are crucial to meet future demands of many civilian, defense and aerospace applications [1-3]. Components (e.g., nozzles, leading edge materials for space vehicles and thermal protection systems) for future high performance aircrafts, hypersonic vehicles, kinetic energy interceptors and reusable space planes, etc., operate in severe reactive environments with temperatures well above the melting points of traditional materials. Few materials can withstand such high temperatures and yet provide adequate mechanical strength. Ultra-high temperature materials (UHTMs) are a class of materials that are chemically and physically stable at temperatures above 2000°C and in reactive atmospheres (e.g., monatomic oxygen) [1-5]. A sub-class of these materials is ultra-high temperature ceramics (UHTCs) which are borides, nitrides and carbides of transitional metals (e.g., Ta, Hf, Zr). These ceramic materials (e.g., ZrB_2, ZrC, HfB_2, HfN, HfC) have high melting point above 3200°C [1-3].

Among UHTCs, ZrB_2 and ZrB_2-SiC composites have emerged as promising candidates because of their low density, high oxidation resistance (above 1500°C), excellent creep resistance, thermal shock resistance and ability to retain strength at elevated temperatures [3-5]. SiC addition improves the processibility of ZrB_2 by reducing the maximum temperature and pressure required to densify the material and simultaneously reducing grain growth of the diboride phase [6]. All these characteristics render ZrB_2-SiC composites attractive for elevated temperature high performance aerospace applications.

51

In recent years, numerous investigations have been conducted on processing and oxidation behavior of ZrB_2-SiC composites. However, the mechanical characterization has been limited mainly to determination of static hardness, fracture toughness, Young's modulus, creep resistance, room and high temperature flexural strength [3,7]. Aerospace structures undergo abrasion due to sharp particles impinging during take off, landings and reentry into earth atmosphere etc., and therefore, evaluation of wear characteristics is important. Such studies on ZrB_2 and its composites are limited. Also, residual stresses due to such wear process can lead to subsurface lateral cracking and can eventually result in failure of the structure. Therefore, it is important to measure the residual stress, as a result of abrasion, in ZrB_2-SiC composites.

Indentation and scratch experiments are commonly used to investigate fundamental features of wear phenomenon between two contacting surfaces [8-13]. From the analysis of evolved damage patterns as a function of external load, fundamental understanding of the deformation mechanisms and material removal processes can be obtained. In the present work, a constant load scratch technique was utilized to observe the induced deformation patterns in ZrB_2-SiC composites.

Raman spectroscopy is a popular nondestructive technique for residual stress measurement [14-17]. The wavelength of a Raman peak of a material is sensitive to the presence of mechanical stress (or more precisely the strain). Therefore, processing- or mechanical deformation-induced residual stress can be measured by Raman spectroscopy. Micro-Raman spectroscopy (MRS), due to its high spatial resolution (of the order of 1 μm or even less), is a useful technique to determine the local residual stress [15,16]. The characteristics Raman peaks in strain-free Raman active crystalline materials correspond to the equilibrium positions of the atoms. Thermal processes or mechanical deformations induce residual strain (thus residual stress) which in turn changes the equilibrium spacing of atoms. Since the residual stress could be either tensile or compressive in nature, bond lengths and thus, force constants, either increase or decrease accordingly compared to the equilibrium values [14,15]. As a result, Raman peak position shifts to a lower or a higher frequency for tensile or compressive residual stress, respectively. From the changes in the Raman peak positions compared to their equilibrium positions, it is possible to measure residual stress present within the material employing Raman spectroscopy. In this paper, preliminary results of residual stress measurements employing MRS will be presented. Although ZrB_2 is not Raman active, SiC is known to be Raman active [18] and hence Raman spectra were collected from the SiC particles within the scratch grooves of ZrB_2-SiC composite.

2. EXPERIMENTAL

2.1. Processing of ZrB_2-SiC composite

Plasma pressure compaction ($P^2C^{\circledR}$) technique has evolved as a promising non-conventational powder consolidation method for sintering of metallic, ceramic and intermetallic particles [19-21]. Unlike the traditional sintering techniques where consolidation times are on the order of several hours, $P^2C^{\circledR}$ method has been successfully employed to sinter materials to full density in few minutes. In $P^2C^{\circledR}$ method, powders are first lightly compressed in a graphite die to establish a current path and then a pulsed DC voltage is applied for surface activation and cleaning of surface contaminants. Upon application of a continuous DC voltage, "Joule" heating occurs at the inter-particle contact areas resulting in softening of the particle contact areas. As the external pressure is increased, rapid densification, over a consolidation period of 5-10 mins, is facilitated through micro-welding and plastic yielding mechanisms. Surface purification and rapid consolidation result in a dense microstructure with minimum grain growth and contamination [19-21].

In the current study, a ZrB_2-5wt%SiC composite, in the form of a rectangular tile of dimensions 75mm×52mm×6mm (see Fig. 1), was processed utilizing $P^2C^{\circledR}$ technique from commercially available

ZrB$_2$ (Grade HS, H.C. Starck, Germany, with particles of sizes 3-7 μm) and polycarbosilane (PCS) powders. PCS is used as a preceramic precursor for SiC which decomposes to SiC upon heating in non-oxidizing atmosphere [22]. Initially, a uniform blend of both the powders was prepared in a shear mixture followed by the calcination of the blend above 1000°C in an atmosphere of flowing argon to decompose the PCS into SiC. Sintering of the heat-treated mixture of ZrB$_2$ and PCS was conducted using the P^{2}C$^®$ method at a temperature of 1750°C and at a pressure of 43 MPa over a short consolidation period of only 5 min. X-ray diffraction (XRD) analysis of the sintered compact confirmed the presence of only two crystalline phases, hexagonal ZrB$_2$ and cubic SiC. Density of the sintered compact, measured by Archimedes method, was found to be above 96% of the theoretical density. Scanning electron micrographs of the fragmented surfaces, see Fig. 2, revealed that the composite was well sintered and thus supported the density measurements.

2.2. Scratch experiments

Rectangular specimens of dimensions 6mm×3mm×4mm were used for scratch experiments. One of the 6mm×3mm surfaces was first ground successively with 120, 240, 320, 400 and 600 grit silicon carbide papers for around 10 min each and then polished for 15 min using 6 μm diamond paste. Scratch experiments were conducted at constant loads of 50, 100, 150, 200 and 250 mN in a MTS nanoindenter$^®$ XPS system employing a Berkovich nanoindenter (tip radius ~ 100 nm). Each of the scratches was conducted for a length of 200 μm at a translational speed of 5 μm/s. Scanning electron microscopy was performed to reveal the microstructural features within the scratch grooves.

2.3. Raman spectroscopy for mechanical residual stress measurements

A Renishaw inVia Raman Microscope was utilized for scratch induced residual mechanical stress measurements. Using a Si-laser (532 nm), Raman spectra were collected from several SiC particles present within the scratch grooves. The laser spot size used was around 1 μm. For comparison purposes, Raman spectra were also collected from several SiC particles residing outside the scratch grooves. All the Raman measurements were performed at room temperature. The Raman spectrometer was calibrated with a Si standard using a Si band position at 520.3 cm^{-1}.

Figure 1. A ZrB$_2$-5wt%SiC composite sintered using P^{2}C$^®$ technique.

Figure 2. SEM micrograph of the fragmented surface of ZrB₂-SiC composite.

3. RESULTS AND DISCUSSION
3.1. Scratch induced deformation and damage patterns

Residual scratch grooves on the polished surface of the ZrB₂-5wt%SiC composite (here after referred as ZrB₂-SiC composite) at three different constant loads of 50, 150 and 250 mN have been shown in Fig. 3. For each of the grooves, only the central portion (~ 80 μm) of the scratch is shown. At lower load levels (50-150 mN), the residual scratch grooves were smooth without any evidence of macroscopic damage (see Figs. 3(a) and 3(b)). But as the load increased, damage appeared along the scratch grooves, see Fig. 3(c). Obviously, the extent of damage increased with load. Figure 4 shows the scratch-length vs penetration-depth profiles at 50, 100, 150, 200 and 250 mN constant scratch loads. Clearly, the scratch depth increased with load. The depth profiles at 50 and 100 mN were relatively smoother compared to those at higher loads (150-250 mN). Greater fluctuations in the scratch depth at higher load levels indicate more damage and material removal during the scratch process, consistent with the micrographs shown in Fig. 3. Similar to the penetration depth profile, Fig. 5 shows the increase in measured scratch grove width with scratch load.

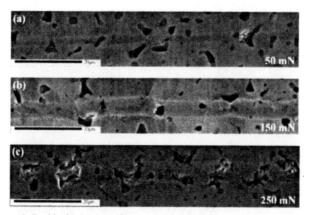

Figure 3. Residual scratch profiles on the polished surface of ZrB_2-SiC composite
at: (a) 50 mN, (b) 150 mN and (c) 250 mN.

Microstructural analysis of the scratch grooves revealed two types of inelastic deformation features: (i) closely spaced parallel lines and (ii) microcracks present mostly at higher loads. Figure 6(a) reveals a scratch pattern at 50 mN load where some regions within the scratch groove, (A) and (B), contained closely spaced parallel lines as shown in Figs. 6(b) and 6(c). The observed closely spaced parallel lines are similar to the 'slip line' patterns observed in ZrB_2 single crystal [23] and indicative of the occurrence of plastic deformation during the scratch process. Also, the slip line patterns were observed only within the ZrB_2 phase but not in SiC particles. In ZrB_2 single crystals, with a hexagonal closed pack (hcp) crystal structure, traces of slip lines were observed during room temperature micro-indentation and high temperature uniaxial experiments due to activation of different set of slip systems [23]. Therefore, it was inferred that in the current investigation, the observed slip line patterns in ZrB_2 phase resulted from the activation of slip mechanism through dislocation motion during scratch experiments.

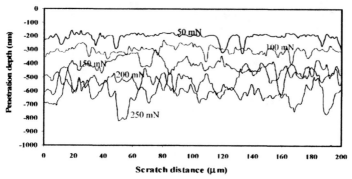

Figure 4. Scratch-length vs penetration-depth profiles at five different constant loads.

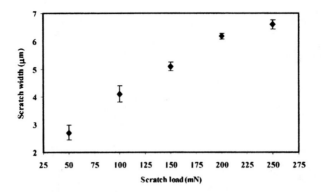

Figure 5. Measured scratch width as a function of scratch load.

It can be clearly seen from Figs. 6(b) and 6(c) that each set of slip lines were oriented randomly with respect to the scratch direction (shown by the white thick arrow). The random orientation of the slip lines in relation to the scratch direction were attributed to the occurrence of dislocation motion along preferred crystallographic slip directions in randomly oriented grains. Changes in slip line orientations across the grain boundary (see Fig. 6(b)) also support this argument. Also, at least two sets of mutually intersected parallel slip lines along the scratch groove could be observed in Fig. 6(b) and Fig. 6(c). Similar deformation features were also observed at higher loads. Figure 7 shows a residual scratch groove at 250 mN load which reveals several sets of slip lines oriented randomly to the scratch direction.

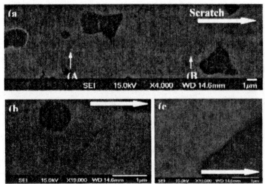

Figure 6.(a) SEM micrograph of scratch patterns at a load of 50 mN. A closer view of the regions (A) and (B) are shown in Fig. 6(b) and Fig. 6(c), respectively, revealing the deformation patterns along the scratch groove.

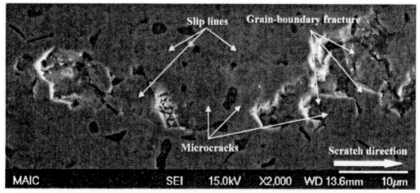

Figure 7. Slip lines and microcracking within a scratch groove at 250 mN.

In addition to slip line patterns, the damaged regions along a scratch groove at 250 mN, see Fig. 7, revealed ZrB_2 grain-boundary fracture and microcracking within the ZrB_2 grains. Microcracking within ZrB_2 grains indicates the transgranular fracture within the ZrB_2 matrix. Clearly, the material removal mechanisms within the scratch groove of the ZrB_2-SiC composite, particularly at higher loads, were a combination of transgranular and grain-boundary modes of fracture within the ZrB_2 matrix. It was interesting to note that the microcracks were present orthogonal to the scratch direction as can be seen clearly from Fig. 8 revealing the damage patterns within a residual scratch groove at 250 mN. Similar orthogonal microcracking can also be seen in Fig. 7.

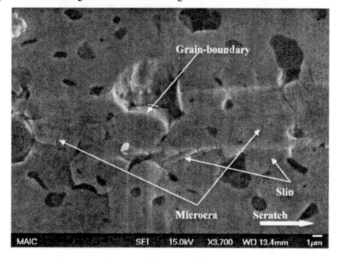

Figure 8. Scratch groove at 250 mN revealing microcracks orthogonal to the scratch direction.

3.2 Influence of elastic stress field on microcracking and slip band formation

The single-pass scratch process has been well studied in the literature and modeled as a sliding microindentation event [10,12,13]. During a scratch process, several stress components develop as a result of elastic-plastic deformation beneath the scratch tool [10,12,13]. In the current investigation, it was speculated that the occurrence of transgranular microcracking orthogonal to the scratch direction and formation of slip bands at an angle to the scratch direction were influenced by the tensile stress component and the shear stress component, respectively. In the following, we will utilize the elastic stress field components to rationalize the formation of slip patterns and microcracking in ZrB_2-SiC composite.

During a scratch process, point loads are applied simultaneously in normal (F_n) and tangential (F_t) directions on the surface of a specimen as shown schematically in Fig. 9. As the indenter moves on the specimen surface, it is assumed that a semi-cylindrical scratch groove of radius a is created which is surrounded by a semi-cylindrical inelastic zone of radius b as shown schematically in Fig. 9. Ahn *et al.*, [10] modeled the scratch process as a sliding indentation event, combining the Boussinesq field due to (F_n), Cerruti field due (F_t) [13,24] and the blister field solutions [25], and rationalized various crack systems which are operative during the scratch event. Recently, Jing *et al.*, [13] estimated the plastic zone size as well as the damage zone size during a scratch process based on a wedge indentation model. Subhash *et al.*, [12] studied the influence of Boussinesq and Cerruti stress fields in the presence of a pre-existing microcrack and identified the most favorable orientation of cracks that activate during a scratch process.

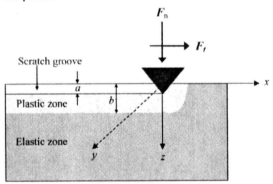

Figure 9. A schematic of the sliding microindentation event with the point normal and tangential loads as well as a moving coordinate system.

In the current investigation, the elastic stress field was constructed from the superposition of Boussinesq field and Cerruti field [13,24]. The Boussinesq field [13,24] results from the application of the point normal load (F_n) and its components on a x-y plane at a depth $z = c$ below the surface are σ_x^n, σ_y^n and τ_{xy}^n. On the other hand, stress components, σ_x^t, σ_y^t and τ_{xy}^t, due to the point tangential load (F_t) are obtained from Cerruti field [13,24]. Here, σ and τ are the normal and the shear stress components and the superscripts n and t refer to the normal and tangential load directions. Now the total normal (σ_x and σ_y) and shear stress (τ_{xy}) components of the elastic stress field beneath the scratch tool during the scratch process are expressed as

$$\sigma_x = \sigma_x'' + \sigma_x', \quad \sigma_y = \sigma_y'' + \sigma_y' \quad \text{and} \quad \tau_{xy} = \tau_{xy}'' + \tau_{xy}'. \tag{1}$$

The expressions for each of the above are provided in [10,13,26].

To explain the influence of stress components on the formation of microcracking and slip line patterns, observed in the ZrB$_2$-SiC composite, the maximum principal stress component (σ_1) and the maximum shear stress component (τ_{max}) have been computed. Figure 10 shows the distribution of the normalized maximum principal stress component at a depth c below the surface, $\bar{\sigma}_1 = 2\sigma_1 \pi c^2 / F_n$, on the $\bar{x} - \bar{y}$ plane and in the vicinity of the scratch tool during the scratch process. The position of the scratch tool is also indicated in the figure. The stress distribution shows the stress singularity directly beneath the scratch tool and tensile principal stress distribution in the wake of the scratch tool. In brittle materials, it is assumed that mode-I cracking occurs when the maximum principal stress exceeds the fracture stress of the material and a mode-I crack opens orthogonal to the direction of maximum tensile stress. Therefore, it is inferred that the existence of tensile principal stress behind the indenter causes the observed microcracking during the scratch process. The maximum principal stress component behind the scratch tool was observed to be oriented in the direction of the scratch process [26] and thus, resulted the mode-I cracking perpendicular to the scratch direction.

Figure 11 represents the distribution of the normalized maximum shear stress component, $\bar{\tau}_{max} = 2\tau_{max} \pi c^2 / F_n$, in the vicinity of the indenter tip on the $\bar{x} - \bar{y}$ plane. Note that the distribution of the $\bar{\tau}_{max}$ is symmetric about the $\bar{x}$ - axis and has highest values in the regions ahead of the indenter position and slightly away from the $\bar{x}$ - axis. From this maximum shear stress distribution, it is postulated that development of high local shear stress ahead of the indenter tip during the scratch process causes slip band formation in these regions. Clearly, Figs. 10 and 11 reveal that during a scratch process, the regions ahead of the indenter tip are subjected to high shear stresses whereas the same regions experience large tensile stress as the indenter moves forward.

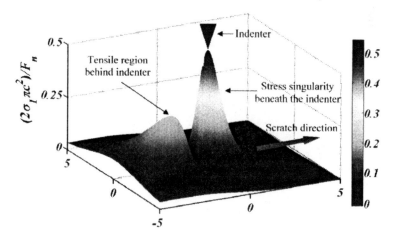

Figure 10. Distribution of normalized maximum principal stress in the vicinity of indenter tip.

The region near the exit-end of each of the scratches was observed in SEM to validate the above analytical result. These end-regions ahead of the scratch tool were only subjected to large τ_{max}, shown in Fig. 11, and were never subjected to high tensile stress as shown in Fig. 10. As a result, only slip lines are expected to develop in these regions. Figure 12 reveals a SEM micrograph of a region at the scratch-end at 250 mN revealing only slip lines present on both sides of the scratch. This observation validates that slip bands formed ahead of the scratch tool due to the occurrence of highest maximum shear stress and supports the above postulation.

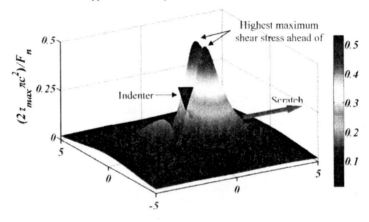

Figure 11. Distribution of normalized maximum shear stress in the vicinity of indenter tip.

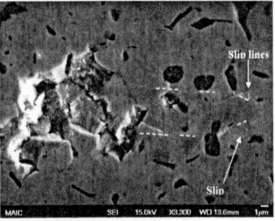

Figure 12: SEM micrograph of the exit-end of a scratch at 250 mN revealing numerous slip lines. The white dashed lines indicate the boundary of the scratch exit-end.

3.3 Influence of scratch loads on Raman spectra

Figure 13 shows the Raman spectra collected from SiC particles within (at 50, 150 and 250 mN loads) and outside of the scratch grooves. All the Raman spectra are typical of 3C-SiC which has two Raman active phonon vibrations; (i) transverse optical (TO) around 796 cm^{-1} and (ii) longitudinal optical (LO) phonon modes around 972 cm^{-1} [18]. These two peak positions have been consistently reported for many annealed 3C-SiC thin films which could be assumed to be stress-free. However, in the current study, SiC particles residing outside the scratch groves showed both the TO- and LO-peaks at higher wave numbers around 802 and 979 cm^{-1}, respectively. Since these peak positions shifted to higher wave numbers, it was inferred that compressive residual stresses are present within the SiC particles (in the original composite) which could arise due to the mismatch in thermal expansion coefficients (CTE) between hexagonal ZrB$_2$ ($\sim 5.9\times10^{-6}$ K^{-1}) and cubic SiC (3.5×10^{-6} K^{-1}) [3] phases. Since the CTE of polycrystalline ZrB$_2$ is greater than that of 3C-SiC, it is expected that during cooling of the consolidated compact from sintering temperature, compressive and tensile residual stresses were generated within SiC and ZrB$_2$ particles, respectively.

Raman spectra, collected from the SiC particles present within the scratch grooves, showed gradual shifting of the TO- and LO-peak positions to lower wave numbers with increasing scratch load. The vertical dashed lines in Fig. 13 indicate the TO- and LO-peak positions in the original composite. Also, peak widening as well as asymmetry in peaks with increasing scratch load can be seen in Fig. 13. Raman peak shift to lower wave numbers indicates that scratch process induced tensile mechanical residual stress within the SiC particles which increased with scratch load. It is well-known that scratch induced tensile mechanical residual stress results in lateral cracking and material removal in brittle materials [10]. The extent of damage increases with increasing accumulation of residual stress within the materials. The increase in tensile residual stress within SiC particles also indicates an overall increase in the induced-damage within the ZrB$_2$-SiC composite with scratch load. The Raman spectroscopic measurements as reported in the current study, will be utilized for mechanical residual stress measurements within the SiC particles present in the scratch groove. Details of the residual stress calculation based on the Raman spectroscopy will be presented in future publications.

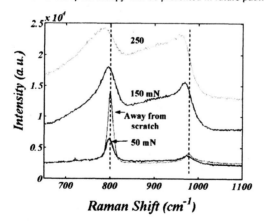

Figure 13. Raman peaks collected from the SiC particles present within and away from the scratch grooves. Note peak shift to lower wave numbers with increasing scratch load

CONCLUSIONS
1. Plasma pressure compaction technique was successfully employed to produce a polycrystalline ZrB_2-5wt%SiC composite, with a sintered density above 96% of theoretical density, at a temperature of 1750°C and at a pressure of 43 MPa over a short consolidation period of only 5 min.
2. Scratch experiments at five different constant loads of 50, 100, 150, 200 and 250 mN using a Berkovich nanoindenter revealed an increase in both the scratch depth and width with scratch load. Scratch grooves at lower scratch load levels (50-150 mN) were smooth without any significant macroscopic damage. But above 150 mN, significant amount of damage and material removal along the scratch path were observed.
3. Microstructural analysis revealed mainly two inelastic deformation features; (i) presence of slip lines oriented at random angles to the scratch direction and (ii) development of microcracks orthogonal to the scratch direction. The slip lines are speculated to occur due to the activation of multiple slip systems or occurrence of dislocation motion along preferred slip systems in randomly oriented grains. Apart from transgranular microcracking, grain-boundary fracture within ZrB_2 phase was also observed.
4. To rationalize the formation of slip lines and orthogonal microcracking, an analytical model, based on the elastic stress field solutions for a single-pass scratch event was utilized. From the mechanistic study, it was postulated that that slip lines are formed during the scratch process due to the development of high shear stress ahead of the scratch tool. As the indenter moves, tensile stress develops in the wake of the indenter causing orthogonal microcracking. Mechanistic analysis also confirmed that during the scratch process, slip lines formed first followed by microcracking.
5. Raman spectroscopy of SiC particles within the scratch groove indicated development of tensile residual stress which increased with increasing scratch load.

ACKNOWLEDGEMENT
This work was funded by a grant from the US NSF (grant # CMS-0324461) with Dr. Ken Chong as the program manager. The authors acknowledge Materials Modifications Inc. (MMI), Fairfax, Virginia, for providing the ZrB_2-5wt%SiC composite. Authors also acknowledge the "Major Analytical Instrument Center (MAIC)", at University of Florida for the use of SEM.

REFERENCES
[1] NSF-AFOSR Joint Workshop on "Future Ultra-High Temperature Materials," *Draft Workshop Report, National Science Foundation*, April 12 (2004).
[2] R Loehman, E Corral, H. P. Dumm, P Kotula and R Tandon, "Ultra-High Temperature Ceramics for Hypersonic Vehicle Applications," *Sandia Report, Sandia national Laboratories* (SAND 2006-2925).
[3] W. G. Fahrenholtz, G. E. Hilmas, I. G. Talmy and J. A. Zaykoski, "Refractory Diborides of Zirconium and Hafnium," *J. Am. Ceram. Soc.*, 90 (5), 1347-1364 (2007).
[4] J. Han, P. Hu, X. Zhang and S. Meng, "Oxidation Behavior of Zirconium Diboride–Silicon Carbide at 1800°C," *Scripta Mater.*, 57, 825–828 (2007).
[5] A. Rezaie, W. G. Fahrenholtz and G. E. Hilmas, "Evolution of Structure During the Oxidation of Zirconium Diboride-Silicon Carbide in Air up to 1500°C," *J. Eur. Ceram. Soc.*, 27, 2495-2501 (2007).
[6] A. Rezaie, W. G. Fahrenholtz and G. E. Hilmas, "Effect of Hot Pressing Time and Temperature on the Microstructure and Mechanical Properties of ZrB_2-SiC," *J. Mater. Sci.*, 42, 2735-2744 (2007).
[7] A. Chamberlain, W. Fahrenholtz, G. Hilmas and D. Ellerby, "Oxidation of ZrB_2-SiC Ceramics Under Atmospheric and Reentry Conditions," *Refract. Appl. Trans.*, 1 (2) 1-8 (2005).
[8] S. Jahanmir, "Friction and Wear of Ceramics," *Mercel Dekker Inc.* (1994).

[9] J. D. Gates and R. A. Etan, "Real Life Wear Processes," *Mater. Forum.*, 17, 369-381 (1993).

[10] Y. Ahn, T. N. Farris and S. Chandrasekar, "Sliding Microindentation Fracture of Brittle Materials: role of Elastic Stress Fields," *Mech. Mater.* 29 (11), 143-152 (1998).

[11] G. Subhash and R. Bandyo, "A New Scratch Resistance Measure for Structural Ceramics" *J. Am. Ceram. Soc.* 88 (4), 918-925 (2005).

[12] G. Subhash, M. A. Marszalek and S. Maiti, "Sensitivity of Scratch Resistance to Grinding Induced Damage Anisotropy in Silicon Nitride" *J. Am. Ceram. Soc.*, 89 (8), 2528-2536 (2006).

[13] X. Jing, G. Subhash and S. Maiti, "A New Analytical Model for Estimation of Scratch Induced Damage in Brittle Solids" *J. Am. Ceram. Soc.*, 90 (3), 885-892 (2007).

[14] E. Anastassakis, A. Pinczuk, E. Burstein, F. H. Pollak and M. Cardona, *Solid State Commun.*, 8, 133-138 (1970).

[15] Y. Kanga, Y. Qiu, Z. Lei and M. Hu, "An Application of Raman Spectroscopy on the Measurement of Residual Stress in Porous Silicon," *Optics & Laser Engg.*, 43, 847–855 (2005).

[16] N. Orlovskaya, D. Steinmetz, S. Yarmolenko, D. Pai, J. Sankar and J. Goodenough, "Detection of Temperature- and Stress-Induced Modifications of $LaCoO_3$ by Micro-Raman Spectroscopy," *Phys. Rev. B*, 72, 14122-1 - 14122-2 (2005).

[17] J.-K. Shin, C. S. Lee, K.-R. Lee and K. Y. Eun, "Effect of Residual Stress on the Raman-Spectrum Analysis of Tetrahedral Amorphous Carbon Films," *APPL. PHYS. LETT.*, 78(5), 631-633 (2001).

[18] S. Rohmfeld, M. Hundhausen and L. Ley, "Raman Scattering in Polycrystalline 3C-SiC: Influence of Stacking Faults," *Phys. Rev. B*, 88 (15), 9858-9862 (1998).

[19] D. Ghosh, G. Subhash, T. S. Sudarshan, R. Radhakrishnan and X. L. Gao, "Dynamic Indentation Response of Fine-Grained Boron Carbide," *J. Am. Ceram. Soc.*, 90 (6), 1850-1857 (2007).

[20] B. R. Klotz, K. C. Cho and R. J. Dowding, "Sintering Aids in the Consolidation of Boron Carbide (B_4C) by the Plasma Pressure Compaction (P^2C) Method," *Mater. Manufac. Process*, 19, 631–639 (2004).

[21] D. P. Harvey II, R. Kalyanaraman and T. S. Sudarshan, "Consolidation and Mechanical Behaviour of Nanocrystalline Iron Powder," *Mater. Sci. Tech.*, 18, 959–963 (2002).

[22] X.-J. Zhou, G.-J. Zhang, Y.-G. Li, Y.-M. Kan and P.-L. Wang, "Hot Pressed ZrB_2-SiC-C Ultra High Temperature Ceramics with Polycarbosilane as a Precursor," *Mat. Lett.*, 61, 960-963 (2007).

[23] J. S. Haggerty and D. W. Lee, "Plastic Deformation of ZrB_2 Single crystals," *J. Am. Ceram. Soc.*, 54 (11), 572-576 (1971).

[24] K. L. Johnson, *Contact Mechanics*. Cambridge University Press, Cambridge, 1985.

[25] E. H. Yoffe, "Elastic Stress Fields Caused by Indenting Brittle Materials," *Phil. Mag. A*, 46 (4), 617-628 (1982).

[26] D. Ghosh, G. Subhash, R. Radhakrishnan and T. S. Sudarshan, "Scratch-Induced Microplasticity and Microcracking in Zirconium Diboride-Silicon Carbide Composite," (Submitted, 2008).

FINITE ELEMENT MODELING OF INTERNAL STRESS FACTORS FOR ZrB_2 – SiC CERAMICS

Michael P. Teague, Gregory E. Hilmas, and William G. Fahrenholtz
Department of Materials Science and Engineering, Missouri University of Science and Technology
Rolla, Missouri

ABSTRACT
Commercial finite element modeling software (ABAQUS and OOF2) was used to calculate the internal stresses produced during cooling after hot pressing of ZrB_2 – SiC ceramics. The size and shape of the SiC inclusions was varied to determine their effect on the residual stresses. Alternate SiC particle shapes designed to limit residual tensile stresses were modeled to minimize residual thermal stresses. Models were validated by measuring actual residual stresses by neutron diffraction. Due to the high neutron absorption of the more common boron isotopes, ZrB_2-SiC parts were fabricated by reactive hot pressing using isotopically pure [11]Boron for these experiments. Characterization of the [11]B materials confirmed similar microstructure and material properties (hardness, Young's modulus and flexure strength) to parts made using powders containing naturally occurring boron. While exact residual stresses for the ZrB_2-SiC composites have yet to be determined, models to investigate the trends in SiC particle size and shape effects, and a method to create neutron diffraction compatible materials, have been accomplished.

INTRODUCTION
Materials for thermal protection systems for hypersonic flight and re-entry vehicles were the focus of research in the past (1960's to 70's) and again in more recent years (2000 to present). Zirconium diboride (ZrB_2), from the family of ultra-high temperature materials, has been studied for use in applications that require high thermal and chemical resistance.[1] ZrB_2 has a high melting temperature (3250°C), high hardness (23 GPa) and a high strength (>500 MPa).[2] Silicon carbide (SiC) is commonly added to ZrB_2 to reduce processing temperatures, control grain size, and improve the mechanical properties and oxidation resistance.[3] A wide variety of processing methods have been used to synthesize these materials such as hot pressing, reactive hot pressing, and pressureless sintering. Reactive hot pressing has shown promise due to the reduced processing temperatures and for the ability to use phase pure starting components such as Zr and Si instead of using commercially available ZrB_2 and SiC powders.[4-6] Equations 1-3 list some of the reactions that have been used to synthesize ZrB_2-SiC composites.

$$2Zr + Si + B_4C \rightarrow 2ZrB_2 + SiC \qquad (1)$$

$$2ZrH_2 + B_4C + Si \rightarrow 2ZrB_2 + SiC + 2H_2 \qquad (2)$$

$$Zr + B + SiC \rightarrow ZrB_2 + SiC \qquad (3)$$

The addition of up to 30 vol% SiC has been shown to markedly improve the strength of ZrB_2-SiC over monolithic ZrB_2 (Table I). Further, the addition of ultra fine (submicron) SiC particles improved sinterability in ZrB_2-SiC composites,[7,8] while providing improved fracture toughness and flexural strength. In other studies, SiC particles were determined to be the strength limiting flaw, and, thus, decreasing SiC particle size resulted in an increase in the strength of ZrB_2-SiC ceramics.[9,10]

Table I. RT Mechanical Properties of ZrB$_2$-SiC Ceramics[2]

SiC Content (vol %)	Modulus (GPa)	Hardness (GPa)	Strength (MPa)	Toughness (MPa*m1/2)
0	489	23 ± 0.9	565 ± 53	3.5 ± 0.3
10	450	24 ± 0.9	713 ± 48	4.1 ± 0.3
20	466	24 ± 2.8	1003 ± 94	4.4 ± 0.2
30	484	24 ± 0.7	1089 ± 152	5.3 ± 0.5

A side effect of the use of a second phase with markedly different linear coefficients of thermal expansion (CTEs) is the residual stresses that are generated during cooling from the processing temperature. Residual stresses in composite materials, and how they impact mechanical properties, have become an increasingly important subject in materials research in recent years.[11] As researchers continue to push materials closer to their property limits, it has become more important to understand and control failure. Residual stresses are important because, when combined with applied stresses, they can lead to premature structural failure. Thus, the ability to limit residual stresses could improve the thermo-structural capabilities of a material. These stresses in multi-phase materials such as ZrB$_2$-SiC, arise from the difference in the CTE of the phases (for ZrB$_2$ CTE = 6.7 x 10^{-6} K^{-1}; for SiC CTE = 4.7 x 10^{-6} K^{-1}).

To attempt to accommodate residual stresses in engineering designs, several modeling and measurement techniques have been investigated. The majority of the modeling techniques used to predict residual stresses fall into two main categories, Eshelby-type and finite element models. Analytical models derived from Eshelby type analysis can be used to predict the stresses using simple mathematical relations, but are often only valid for mean phase stresses.[12,13] Finite element models can be used to model complex constitutive laws, but are often limited by idealist unit cell representations.[14] Finite element analysis (FEA) programs have the ability to run the complicated equations on microstructures built with computer aided design (CAD) software. Programs such as ABAQUS[15] and Object Oriented Finite Element Analysis (OOF2)[16] have proven useful for this type of analysis.[17-19] Other programs have also been used to model thermal residual stresses, and in some cases to attempt to relate those stresses back to strength and fracture toughness.[20-24]

Experimental techniques for measuring residual stresses in materials can be used to validate the models. The primary non-destructive methods for determining residual stresses are X-ray and neutron diffraction. X-ray diffraction is the least expensive and more widely available testing method; however, due to its low depth of penetration (< 50 μm) it is only effective for measuring residual stresses near external surfaces.[25] Neutron diffraction, in contrast, is more promising because of improved penetration due to a wavelength closer to atomic spacing.[26-28] Use of neutron diffraction has been limited by cost and access to pulsed neutron sources. Another limiting factor for neutron diffraction is the neutron absorption characteristics of the elements being tested. The thermal neutron absorption for a typical, commercial grade boron (B) is around 770 barns, based on the volumetric contributions of the boron isotopes (20 vol% ^{10}B, 3835 barns; 80vol% ^{11}B, 0.0055 barns). A neutron absorption of ~100 barns is too high for use in a neutron diffraction experiment.[29] A high absorption cross section results in low scattered intensity, which is detrimental to neutron diffraction analysis. Detectors in the system pick up scattered neutrons to create diffraction patterns, so a higher scattering ability (low neutron absorption) is required. Therefore, isotope ^{11}B was used in the current study, along with Zr and SiC, to fabricate ZrB$_2$-SiC materials for residual stress testing.

The objective of this study was to model residual stresses in ZrB$_2$-SiC ceramics that result from the CTE mismatch. SiC particle size and shape were varied to determine their impact on the resultant stresses. ZrB$_2$-SiC parts produced using the ^{11}B isotope were fabricated so that the residual stresses could be measured by neutron diffraction analysis to validate the models.

EXERIMENTAL PROCEDURE AND MODEL DEVELOPMENT

Model Creation and Setup

Using a commercially available finite element analysis program, ABAQUS, models representing SiC particulates in a ZrB$_2$ matrix were created. The SiC phase was modeled as a round particle in a two dimensional (2D) ZrB$_2$ matrix. Material properties were assumed to be isotropic for both the ZrB$_2$ and the SiC after initial modeling efforts indicated only small changes in stress fields as a result of the anisotropic properties of the α-SiC (hexagonal polytype). The material properties used in the models, as well as other key model input variables, are included as Table II.

Table II. Material Properties and Modeling Parameters

Material Properties	ZrB$_2$	SiC	Modeling Parameters	
Young's Modulus[33] (GPa)	489	430	Surface Film Coefficient	10 W/m^2 • K
Thermal Conductivity[33] (W/m•K)	64.4	110	Sink Temperature	25°C
CTE[33] (μm/μm•K)	6.80 x 10^{-6}	4.50 x 10^{-6}	Initial Temperature	1900°C
Density[2] (g/cm^3)	6.26	3.18	Mesh Seeding Size	0.15 μm
Specific Heat[33] (J/g•K)	0.5	0.67	Step Size	0.004 sec
Poisson's Ratio[33]	0.15	0.17	Time Period	108000 sec

The first step was to build a 2D matrix measuring 15μm square. The SiC particles were set on a grid system to limit random interaction and overlap and were from 4 (2x2 grid) to 400 (20x20 grid) in number. The phases were sized such that there was always ~30 vol % SiC in the composite. These particles were assigned to a section with SiC material properties, while the matrix was assigned another section with ZrB$_2$ properties. Next, boundary conditions were applied to simulate the heating and cooling conditions that are typical of the hot pressing cycle for ZrB$_2$-SiC ceramics.[2] To accomplish this, the part had to be heated without creating any stress or expansion. This was accomplished by using a field condition that set the initial temperature without altering the microstructure in any way. The cooling boundary condition was setup using a surface film condition. The surface film condition cools the part according to equation 4,

$$q = -h \times (T - T_o) \tag{4}$$

where q is the heat flux, h is the surface film coefficient and T_o is the temperature of the surrounding material, which is also known as the sink temperature. The sink temperature allowed for the cooling rate of all four sides to be adjusted individually to account for the location of the model within a larger part. Because the models were a local/global approximation, it was important that the model be able to predict different behavior based on the sample being either completely within a larger part or located near the edge of a billet.

With all the material properties and boundary conditions set, the part was then meshed. A coupled temperature-displacement element type was used, which allowed for static stress/strain relationships to be run along with thermal stress/strain and heat fluxes. The model was given a step size of 0.004 seconds with a maximum allowable temperature change per step of 50°C. This created steps that allowed examination of the stresses at nearly any temperature during the cooling process. Post processing was performed using the system available in ABAQUS and was used to determine the stresses in the material.

Modeling Size and Shape Effects

With the modeling process, an effort was made to model ZrB$_2$-SiC ceramics with different microstructures. Shape effects were explored by modeling different SiC shapes and sizes using the same grid pattern used in the earlier size effect models (Table III). Squares, hexagons, and peanut shapes were all modeled to further explore the size effects and any effect the shape of the SiC particle might have on the residual stresses. The peanut shape was created in an attempt to minimize the addition of stresses from neighboring particles. The particles were aligned such that each particle was rotated 90° relative to its neighbors such that a convex side on one particle always faced a concave side of its neighbors.

A spiral shaped particle was also explored. Based on a spiral shape created in earlier research in Al$_2$O$_3$ – ZrO$_2$ systems, SiC particles were formed into spiral shapes allowing ZrB$_2$ to penetrate between each layer of the spiral as shown in Figure 1.

Table III. Sizes of SiC particles for shape models

Grid	Hexagon Leg (μm)	Square Side (μm)	Circle Diameter (μm)	Peanut Radius (μm)
2x2	2.549	4.108	4.635	1.554
3x3	1.699	2.739	3.090	1.036
4x4	1.274	2.054	2.318	0.777
5x5	1.019	1.643	1.854	0.621
6x6	0.850	1.369	1.545	0.518
7x7	0.728	1.174	1.324	0.444
8x8	0.637	1.027	1.159	0.389
9x9	0.566	0.913	1.030	0.345
10x10	0.510	0.822	0.927	0.311

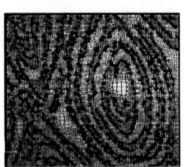

Figure 1. Spiral architecture meshed by OOF2. Red highlights SiC section.

OOF2

In addition to ABAQUS, the program Object Oriented Finite Element Analysis (OOF2; freeware available from the National Institute for Standards and Technology) was used to model an actual ZrB$_2$-SiC microstructure from micrograph obtained by scanning electron microscopy (SEM). The micrograph was imported into the program, and using image enhancement tools, was enhanced such that the contrast between the darker SiC phase and the lighter ZrB$_2$ phase was a large as possible. OOF2's image selection tools were then used to select all of the SiC using a combination of the 'burn method,' which selects other equivalent neighboring colors according to set contrast values, and individual pixel selection to improve boundaries between the materials. Once selected, the pixels representing the SiC and ZrB$_2$ were assigned to groups and given material properties.

The most complicated process in OOF2 was the creation of the skeleton. The skeleton is the backbone by which the mesh is created and the boundary conditions are assigned. Starting with four square elements, the skeleton was refined until the homogeneity of the elements was better than 99%. Initially the elements contained some SiC pixels and some ZrB$_2$ pixels, resulting in poor homogeneity (approximately 40%). The program provided extensive tools to split the initial elements into smaller elements, move them along boundary conditions, and to pin them in certain locations in order to obtain the best possible skeleton. Individual elements were also selected on the basis of their homogeneity and modified to create a finer mesh at the boundaries while allowing for larger elements within larger sections of one material or the other. Once completed, the program created a mesh over top of the skeleton. Boundary conditions could then be applied using OOF2, but to maintain the original modeling conditions, the skeleton, mesh, and material sections were exported into ABAQUS using a program called OOF2ABAQUS. This allowed the model created in OOF2 to have the exact same boundary conditions applied as the other models and the same processor and post-processing to be run on the OOF2-based model. Using OOF2 it was then possible to compare idealized models to actual microstructures.

Powder Processing

ZrB$_2$-SiC composites were fabricated using the [11]Boron ([11]B) isotope so the residual stresses could be measured by neutron diffraction. Crystalline [11]B (Isotopically 99.65 At% [11]B, >99.99 wt% phase pure) with an average starting particle size of around 10μm was donated by EaglePicher, LLC. The boron powder was ball milled in hexane using tungsten carbide media for 17 hours to reduce the particle size to ~1μm. The powder was separated from the hexane by rotary evaporation (Model Rotavapor R-124, Buchi, Flawil, Germany) at a temperature of 70°C, a vacuum of ~30kPa, and a rotation speed of 140 rpm. The [11]B powder was then combined with other commercial powders ZrH$_2$ (Grade C, Chemetall, Frankfurt, Germany) and SiC (UF-25, H.C. Starck, Newton, MA) to produce a ZrB$_2$-30 vol% SiC composite according to equation 5, and based on previous reaction processing studies.[3-5]

$$ZrH_2 + {}^{11}B_2 + SiC \rightarrow Zr^{11}B_2 + SiC + H_2(g) \tag{5}$$

After batching, the powder was attrition milled (Model HD-01; Union Process, Akron, OH) in hexane at 600 rpm for 240 min in a fluoropolymer coated bucket using tungsten carbide media and a tungsten carbide spindle to reduce the starting particle sizes. The powder was again dried by rotary evaporation.

The composites were then densified using a reaction hot pressing process. The powder was loaded into a graphite die lined with BN-coated graphite foil, and was compacted at a 30 MPa uniaxial pressure. The temperature of the graphite die was monitored using an infrared pyrometer (Model OS 3708, Omega Engineering, Stamford, CT). From room temperature to 1650°C, the furnace was heated under a vacuum of ~20 Pa. At 1650°C the chamber was backfilled with argon and a flowing argon atmosphere was maintained through the rest of the hot pressing cycle. The furnace was heated at a rate of ~5°C/min up to 1000°C, the temperature at which all of the H$_2$ gas had been released, as indicated by a decreased vacuum pressure. After release of the H$_2$, the heating rate was increase to ~20°C/min. When the furnace reached 1900°C, a uniaxial pressure of 32 MPa was applied for 45 minutes after which the furnace was turned off and allowed to cool. The pressure was released at 1700°C. The resulting billets were nominally 4.625 mm in diameter x 3 mm thick after hot pressing.

Characterization
    Several characterization experiments were performed to compare the composite made with [11]B
to those produced with natural B consisting of mixed B isotopes. The microstructures of polished
cross sections were examined using a scanning electron microscope (SEM; Model S4700 Hitachi,
Tokyo, Japan). The bulk density of each part was measured using the Archimedes method with
distilled water as the immersing medium. Vickers' microhardness measurements (Model V-1000-A2,
Leco, St. Joseph, MI) were preformed using a test load of 2 kg and a dwell time of 10s. Flexural
strength was determined using a 4-point bend test according to ASTM standard C1161-02a using size
A bend bars (nominally 1.5 mm by 2 mm by 25 mm). According to ASTM Standard C1259-01,
Young's modulus was measured by impulse excitation (Model MK4-I Grindasonic, J.W. Lemmens, St.
Louis, MO). X-ray diffraction (XRD: Scintag, XDS 2000, Cupertino, CA) analysis was used to
confirm the phase composition of the material after reaction hot pressing.

Neutron Diffraction
    Neutron diffraction was performed on three samples to determine residual stresses at the
Intense Pulsed Neutron Source (IPNS) facility of Argonne National Laboratory. Using the general
purpose powder diffractometer (GPPD), flexure bars of ZrB₂-SiC were loaded onto the alumina holder
in a molybdenum furnace and placed under vacuum. Data was gathered for 2 hours at temperatures
ranging from room temperature (~25°C) up to 1200°C (using 100°C increments to 700°C, and then
50°C increments to 1200°C). Powder diffraction studies, which are required for calculation of residual
stresses, have not been completed, but will be reported in a subsequent publication.

RESULTS AND DISCUSSION

MODELING EFFORTS

Size Effect
    Finite element models based on ZrB₂- 30 vol % SiC composites containing idealized round SiC
particles, predicted compressive residual stresses in the dispersed SiC particles while the ZrB₂ matrix
immediately surrounding each particle was in tension. Previously an Eshelby analysis shown in
equation 6[30], had been used to predict radial compressive stresses of as high as 2.1 GPa within the ZrB₂
matrix and tangential tensile stresses of 4.2 GPa at the ZrB₂-SiC boundaries.[3]

$$\sigma_{radial} = -2 \times \sigma \tan = \frac{(\alpha_m - \alpha_i)\Delta T}{(1 - 2\nu_i)E_i + (1 - \nu_m)(2E_m)}\left(\frac{R}{r + R}\right)^3 \qquad (6)$$

Equation 6 assumes round SiC inclusions of radius R, and predicts the stress at a distance r from the
interface where m denotes matrix, i denotes inclusion, α represents CTE, ν is Poisson's ratio, and E is
Young's modulus.
    Stress fields were also observed to overlap between particles, resulting in larger tensile stress
fields as shown in Figure 2. The maximum and minimum stresses for each model were recorded and
plotted alongside experimental results for flexural strength vs. SiC particle size (Figure 3). The results
showed that as particle size decreased, the magnitude of the tensile stresses in the matrix decreased,
while the magnitude of the compressive stresses in the particles increased. This corresponded to an
increase in strength as the particle size decreased. While the exact magnitudes of the residual stresses
are unknown, the data demonstrates a strong correlation between the magnitude of the residual stresses
and measured strength. The compressive stresses in the SiC particles, as well as the tensile stresses
surrounding them, are consistent with a higher propensity to deflect cracks. Crack deflection would

force the crack into a different mode and that energy consuming effect would improve the strength of the material, which agrees with findings in previous studies.[7,8] The smaller particles also reduced the size of the areas in the matrix that were under the most severe tensile load, decreasing the probability that a critical flaw would form within those areas.

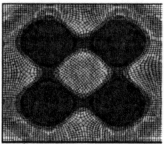

Figure 2. Round SiC particles in a ZrB_2 matrix showing SiC particles in compression and the matrix in tension.

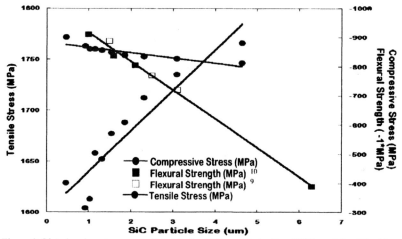

Figure 3. Plot showing stresses predicted in the ZrB_2 matrix (red) and SiC particles (blue) by compared to measured flexure strengths reported in recent literature.

Shape Effect

Models containing 30 vol% SiC, having square or hexagonal shapes, showed a similar trend to the round SiC particles where decreasing SiC particle size decreased the magnitude of the tensile stresses in the matrix and increased the magnitude of the compressive stresses in the inclusions. However, the data overlap made it difficult to determine if one or more of the shapes might be superior in terms of the strength of an as-processed ceramic. One conclusion that can be drawn from the models is that sharp points on the SiC particles (with increasing sharpness going from round to hexagonal to square) increased the residual tensile stresses in the ZrB_2 at the ZrB_2-SiC interfaces. The sharp corners acted as stress concentrators, an effect observed in other, similar modeling efforts.[20] A final result of this initial study was that the square particles showed a reduced combined stress affected area in the matrix. The peanut shape was then developed as a way to enhance this effect. The peanut shape would, theoretically, provide no sharp corners and so that the highest stress point on each particle, the convex ends, would be facing the lowest tensile face of the neighboring particle, which was the concave middle. The models concluded that even though there was a high stress point at the end of each particle, there was little combined stress effect between the particles due to their specific orientations.

After the initial results of the SiC particle studies, the spiral architectures were developed. Early results of spiral shaped SiC inclusions exhibited a smaller stress gradient between the matrix and particles indicating a promising microstructure. While the overall magnitudes of the tensile stresses were not significantly lower than other shapes, it was theorized that a smaller stress gradient would lead to fewer micro cracks during cooling and improve the overall strength of macroscopic parts. Subsequent fabrication of ceramics containing novel SiC particle shapes in a ZrB$_2$ matrix resulted in much lower strengths (~150 MPa) than was expected based on the original models. Returning to the models and using OOF2 to model actual microstructures, it was determined that due to the large size of the SiC inclusions (~100 μm in diameter) the volume affected by the residual tensile stresses was much greater than originally modeled leading to a greater probability of a containing a critical flaw within those areas, which would account for lower strengths.

[11]BORON / NEUTRON DIFFRACTION

A ZrB$_2$-30 vol% SiC composite was fabricated from [11]B, ZrH$_2$, and SiC by reaction hot pressing. Archimedes analysis revealed the parts to be ~99% of their theoretical density, assuming that approximately 1 wt% WC was picked up from the milling media during attrition milling. Comparison of known diffraction patterns for ZrB$_2$[31] and SiC[32] to a neutron diffraction pattern collected from the reaction hot pressed parts confirmed that a ZrB$_2$-SiC composite had formed (Figure 4). Vickers' microhardness results, along with results from strength testing, were compared to values found in the literature to confirm that the material was similar to those produced using powders containing naturally occurring boron (Table IV). Hardness values ranged from 27 GPa (ZrB$_2$-SiC) to 20.5 (Zr[11]B$_2$-SiC) and Young's modulus numbers ranged from 510 (ZrB$_2$-SiC) to 412 (Zr[11]B$_2$-SiC), while the strength numbers were less than half of previous values of 1089 and 800 MPa. Visual comparison of the micrographs for ZrB$_2$-SiC and Zr[11]B$_2$-SiC showed that the [11]B material appeared to have grains roughly twice the size of the higher strength materials. According to previous work,[9,10] larger SiC grains lead to decreased strength, which may explain the lower strength in the materials containing [11]B. However, the strength results are based on a small sample set (3) of flexure bars, so further testing must be conducted to determine an accurate flexure strength value.

Table IV. Comparison of mechanical properties
of Zr[11]B$_2$-SiC to other ZrB$_2$-SiC materials

	HP[2] ZrB$_2$-SiC	RHP[6] ZrB$_2$-SiC	RHP Zr[11]B$_2$-SiC
E (GPa)	484	510	412
Strength (MPa)	1089 ± 152	800 ± 115	373 ± 110
Hardness (GPa)	24 ± 0.7	27 ± 2.2	20.5 ± 0.6

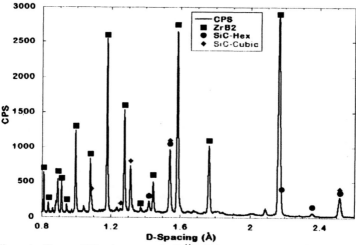

Figure 4. Neutron diffraction pattern for Zr^{11}B$_2$-SiC

Although not all of the experiments needed to determine residual stresses have been completed, some useful data has been gathered. As temperature increased, the diffraction peaks shifted to lower two theta values, indicating an increase in d-spacing correlating to thermal expansion of the composite material.(Figure 5) Once analysis of the powder samples is complete, the inherent thermal expansion of each phase can be calculated, which will allow the strains to be determined for each phase in the composite material. The diffraction patterns from the experiments also provided evidence that the use of ^{11}B to produce the parts made neutron diffraction possible. The patterns were visible within the first few minutes of data acquisition, which indicated the excellent scattering from the material.

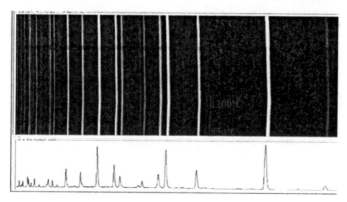

Figure 5. ISAW output showing peak shifts as a function of temperature. The full diffraction pattern for 25°C is displayed at the bottom.[14]

CONCLUSIONS

Using commercially available finite element analysis software (ABAQUS), models were constructed to represent ZrB_2-30 vol.% SiC ceramics. The models, while unverified, provided insight into the nature of the residual stresses that develop during cooling from the final processing temperature as a result of the thermal expansion mismatch between the constituents of the composite. Stress values were compared to strength data taken from the literature and indicate trends that agree with published experimental data. Modeling has also been used to help explain low strengths measured for ZrB_2 containing novel SiC inclusions by focusing on the large areas under high tensile load due to the large size of the SiC inclusions.

Fully dense (~99%) ZrB_2-SiC parts were successfully fabricated by reaction hot pressing of [11]B isotope, ZrH_2 and SiC powders. The material had comparable hardness (20.5 to 27 GPa) and Young's modulus (412 to 510 GPa) to published data, for similar materials made from commercially available powders, respectively. Strength values were less than half of the values reported from previous work, but more complete experiments must be run to confirm the strength of this material. Samples cut from these parts have been tested and successfully examined using the pulsed neutron diffraction sources at Argonne National Laboratory. Subsequent experiments will be used to confirm the magnitude of residual thermal stresses.

ACKNOWLEDGEMENTS

This work was supported by contract FA9550-06-1-0125 with the High Temperature Aerospace Materials Program (Dr. Joan Fuller, program manager) in the Air Force Office of Scientific Research.

The authors would also like to thank EaglePicher, LLC. for the generous donation of [11]B for this research, Shuangmei Zhao of Missouri S&T for extensive support with the Abaqus modeling, Steve Langer at NIST for help installing and learning OOF2, and Joe Fieramosca and Ryoji Kiyanagi at Argonne National Laboratory for running the Neutron Diffraction experiments as well as help with processing the data.

REFERENCES

[1] A.L. Chamberlain, W.G. Fahrenholtz, G.E. Hilmas and D.T. Ellerby, "Characterization of Zirconium Diboride for Thermal Protection Systems," *Key Engineering Materials.* **264-268** 493-496 (2004)

[2] A.L. Chamberlain, W.G. Fahrenholtz, G.E. Hilmas and D.T. Ellerby, "High Strength ZrB_2-Based Ceramics," *J. Am. Ceram. Soc.* **87** [6] 1170-2 (2004)

[3] W. G. Fahrenholtz, G.E. Hilmas, "Refractory Dibordies of Zirconium and Halfnium," *J. Am. Ceram. Soc.,* **90**[5], 1347-1364, (2007).

[4] G.-J. Zhang, Z.-Y. Deng, N Kondo, J.-F. Yang, and T. Ohji, "Reactive Hot Pressing of ZrB_2-SiC Composites," *J. Am. Ceram. Soc.,* **83** [9] 2330-2 (2000).

[5] J.W. Zimmermann, G.E. Hilmas, W.G. Fahrenholtz F. Monteverde, and A. Bellosi, "Fabrication and properties of reactively hot pressed ZrB_2-SiC ceramics," *Journal of the European Ceramic Society,* **27** 2729-2736 (2007).

[6] A.L. Chamberlain, W.G. Fahrenholtz, and G.E. Hilmas, "Low Temperature Densification of Zirconium Diboride Ceramics by Reactive Hot Pressing," *J. Am. Ceram. Soc.* **89** [12] 3638-3645 (2006)

[7] F. Monteverde, A. Bellosi, "Development and characterization of metal-diboride-based composites toughened with ultra-fine SiC particulates," *Solid State Sciences*, 7, 622-630, (2005).

[8] F. Monteverde, "Beneficial effects of an ultra-fine α-SiC incorporation on the sinterability and mechanical properties of ZrB$_2$," *Applied Physics A*, 82, 329-337, (2006).

[9] A. Rezaie, W.G. Fahrenholtz, and G.E. Hilmas, "Effect of hot pressing time and temperature on the microstructure and mechanical properties of ZrB$_2$-SiC," *J. Mater Sci*, 42 2735-27-44 (2007)

[10] S. Zhu, W.G. Fahrenholtz, and G.E. Hilmas, "Influence of SiC particle size on the microstructure and mechanical properties of ZrB$_2$-SiC ceramics," *ECERS*, 27, 2077-2083, (2007).

[11] P.J. Withers and H.K.D.H. Bhadeshia, "Residual Stress, Part 2 – Nature and origins," *Material Science and Technology*, 17, 366-375, (2001).

[12] J.D. Eshelby, "The determination of the elastic field of inclusion and related problems," Proc. R. Soc. Lond., A 241, 376-396. (1957).

[13] P.J. Withers, W.M. Stobbs, and O.B. Pedersen, "The application of the Eshelby method of internal stress determination to short fibre metal matrix composites," *Acta Metall.*, 37, 30061-3084, (1989).

[14] T.W. Clyne and P.J. Withers: in "An Introduction to Metal Matrix Composites", pp. 34-43, Cambridge Solid State Series; Cambridge University Press, Cambridge, England, (1993).

[15] www.abaqus.com

[16] www.ctcms.nist.gov/oof/oof2/

[17] P. Rangaswamy, C.P. Scherer, M.A.M. Bourke, "Experimental measurements and numerical simulation of stress and microstructure in carburized 5120 steel disks," *Materials Science & Engineering A*, 298, 158-165, (2001).

[18] I. Ozdemir and M. Toparli, "An Investigation of Al-SiCp Composites Under Thermal Cycling," *Journal of Composite Materials*, 37[20], 1839-1850, (2003).

[19] A. Saigal, E.R. Fuller, S. Jahanmir, "Modeling of Residual Stresses and Mechanical Behavior of Glass-Infiltrated Spinel Ceramic Composites," *Computational Modelling of Materials, Minerals and Metals Processing*, (ed M. Cross, J.W. Evans and C. Bailey), TMS, (2001).

[20] V.E. Saouma, S.-Y. Chang, O. Sbaizero, "Numerical simulation of thermal residual stress in Mo- and FeAl-toughened Al$_2$O$_3$," *Composites: Part B*, 37, 550-555, (2006).

[21] Y.-W. Bao, C.-C. Liu, J.L. Huang, "Effects of residual stresses on strength and toughness of particle-reinforced TiN/Si$_3$N$_4$ composite: Theoretical investigation and FEM simulation," *Materials Science & Engineering A*, 434, 250-258, (2006).

[22] T.-P. Lin, M. Hood, G.A. Cooper, "Residual Stresses in Polycrystalline Diamond Compacts," *J.Am. Ceram. Soc.*, 77[6], 1562-68, (1994).

[23] P. Rangaswamy, M.B. Prime, M. Daymond, M.A.M. Bourke, B. Clausen, H. Choo, N. Jayaraman, "Comparison of residual strains measured by X-Ray and neutron diffraction in a titanium(Ti-6Al-4V) matrix composite," *Materials Science & Engineering A*, 259, 209-219, (1999).

[24] P. Rangaswamy, M.A.M. Bourke, P.K. Wright, N. Jayaraman, E. Kartzmark, J.A. Roberts, "The influence of thermal-mechanical processing on residual stresses in titanium matrix composites," *Materials Science & Engineering A*, 224, 200-209, (1997).

[25] D. Magley, R.A. Winholtz, K.T. Faber, "Residual Stresses in a Two-Phase Microcracking Ceramic," *J. Am. Ceram. Soc.*, 73[6], 1641-44, (1990).

[26] T. Lorentzen and T. Leffers, "Strain Tensor Meaurements by Neutron Diffraction," in Measurements of residual and applied stress using neutron diffraction, Vol 216, NATO ASI Series E (ed M.T. Hutchings and A.D. Krawitz), 253-261, Dordrecht, Kluwer, (1992).

[27] M.R. Daymond, M.A.M. Bourke, R.V. Dreele, B. Clausen, T. Lorentzen, "Use of Rietveld refinement for elastic macrostrain determination and for evaluation of plastic strain history from diffraction spectra," *J. Appl. Phys.*, 82, 1554-1556, (1997).

[28] M. Nishida, M. Rifaimulish, Y. Ikeuchi, N. Minakawa, T. Hanabusa, "Low temperature in-situ stress measurement of W/Cu composite by neutron diffraction," *International Journal of Modern Physics B*, **20**, 3668-3673, (2006).

[29] A.D. Krawitz, Introduction to Diffraction in Material Science and Engineering, New York: John Wiley & Sons, 2001.

[30] M.W. Barsoum, "Thermal Stress and Thermal Properties," pp. 488-492 in *Fundamentals of Ceramics*, McGraw-Hill Companies, New York, NY, 1997.

[31] Card number 34-0423, Powder Diffraction File, International Centre for Diffraction Data.

[32] Card number 49-1428, Powder Diffraction File, International Centre for Diffraction Data.

[33] R.A. Cutler, "Engineering Properties of Borides", pp 787-803 in *Ceramics and Glasses: Engineered Materials Handbook*, Vol 4, Edited by S.J. Schneider Jr. ASM International, Materials Park, OH, 1991.

[34] http://www.pns.anl.gov/computing/isaw/index.shtml

EFFECTS OF MICROSTRUCTURAL ANISOTROPY ON FRACTURE BEHAVIOR OF HEAT-PRESSED GLASS-CERAMICS AND GLASS-INFILTRATED ALUMINA COMPOSITES FOR DENTAL RESTORATIONS

Humberto N. Yoshimura
Institute for Technological Research of the State of São Paulo
São Paulo, São Paulo, Brazil

Carla C. Gonzaga, Paulo F. Cesar, and Walter G. Miranda Jr.
School of Dentistry, University of São Paulo
São Paulo, São Paulo, Brazil

ABSTRACT
    The effects of microstructural anisotropy associated with sample geometry on fracture behavior were investigated for three dental ceramics (glass-ceramic reinforced by near-isometric leucite particles, Empress - E1, glass-ceramic reinforced by elongated lithium disilicate particles, Empress 2 - E2, and glass-infiltrated alumina composite, In-Ceram Alumina - IC). The glass-ceramic samples were prepared by the heat-pressing method and the IC samples were prepared by infiltration of a La_2O_3 based glass into a skeleton of pre-sintered alumina platelets. The samples were prepared with bar and disc geometries. The material with near-isometric particles (E1) had an isotropic microstructure and K_{Ic} values were statistically similar for both sample geometries. The materials with particles with high aspect ratios (E2 and IC) had anisotropic microstructures and K_{Ic} values of the samples with the bar geometry were significantly higher than the samples with the disc geometry, because of the reinforcement caused by particle alignment along the direction of longer axis of the bar. For sample E2, the heat-pressing method caused particle alignment along the direction perpendicular to the direction of injection in disc specimens, which resulted in relatively low biaxial flexural strength caused by the formation of weak paths in the glassy matrix for crack propagation. For evaluation of the strength of materials which tend to exhibit microstructural anisotropy caused by processing variables, the biaxial flexural test is recommended over uniaxial bending, since it allows for crack propagation by weakest microstructural path.

INTRODUCTION

    The increasing demand for aesthetic materials in dentistry has pushed the development of novel all-ceramic systems. Besides excellent aesthetics, these materials have the advantage of increased strength, color stability, wear and abrasion resistance and biocompatibility.[1] Nowadays, many all-ceramic systems are available for the construction of monolithic (inlays, onlays, overlays and crowns) or bilayered (crowns and fixed partial dentures – FPDs) restorations. Among these systems, heat-pressed glass-ceramics are one of the most popular options because of their ease of fabrication by the lost-wax technique, better marginal fit, good mechanical properties and decreased porosity when compared to conventional feldspathic porcelains.[2,3]

    Glass-ceramics are produced from suitable glasses by a thermal treatment known as controlled crystallization. The bulk composition and the microstructure of glass-ceramics determine the general physical and chemical characteristics of the final material, as well as their mechanical and optical properties.[4] IPS Empress (E1) (Ivoclar, Schaan, Liechtenstein) is a leucite-reinforced heat-pressed glass-ceramic released in the market in the early 1990s and can be used to fabricate single unit restorations, including veneers, inlays, onlays and crowns. This material is based on the SiO_2–Al_2O_3–K_2O system and has the following composition (in wt%): 63% SiO_2, 17.7% Al_2O_3, 11.2% K_2O, 4.6% Na_2O, 0.6% B_2O_3, 1.6% CaO, 0.2% TiO_2, 0.7% BaO, 0.4% CeO_2, plus pigments.[5] Its microstructure consists of evenly dispersed leucite crystals embedded in a glassy matrix,[6,7] with a reported leucite content between 36 and 40% in volume.[7-9] In 1998, the same manufacturer developed another glass-ceramic, IPS Empress 2 (E2), based on the SiO_2–Li_2O system, which contains lithium disilicate as the main crystalline phase. According to the manufacturer, this glass-ceramic allows the fabrication of 3-unit FPDs up to the second premolar region.[10] Its composition consists of the following, in wt%: 57-80% SiO_2, 11-19% Li_2O, 0-13% K_2O, 0-11% P_2O_5, 0-8% ZnO, 0-5% MgO, 0.1-6% La_2O_3, 0-5% Al_2O_3, plus 0-8% pigments.[4] This material has a unique microstructure composed of about 65% in volume of elongated interlocked lithium disilicate crystals dispersed in its glassy matrix.[7,10-12] Both glass-ceramics described above are processed by the heat-press technique. In this processing method, the waxed pattern of the restoration is invested in a refractory material, which is preheated to 850°C for one hour to eliminate wax. The refractory mold was subsequently transferred to a special press furnace. The pre-cerammed ingots, supplied by the manufacturer in a variety of shades, are then placed in the open end of the mold and pressed by a thermally resistant alumina plunger under pressure from the furnace.[6]

Another all-ceramic dental restoration system for construction of 3-unit FPDs is based on the infiltration of glass into porous skeletons of ceramic crystal particles, which results in ceramic-glass composite materials. In 1990, In-Ceram Alumina (IC) (Vita-Zahnfabrik, Germany) prepared by this method was introduced in the market.[13] First, a green preform is prepared by slip casting of Al_2O_3 particles and platelets, which is partially sintered to enhance the strength, but with minimum shrinkage to guarantee good marginal fit. Afterwards, the pre-sintered preform is covered with a lanthanum-aluminum silicate glass powder and then heat-treated to promote spontaneous infiltration.[14-17] The microstructure of this composite has around 68 vol.% alumina, 27 vol.% glass, and 5 vol.% porosity.[7] The same manufacturer also developed two others composites, In-Ceram Spinell (with $MgAl_2O_4$ particles) and In-Ceram Zirconia (with Al_2O_3 and ZrO_2 particles).[18]

The mechanical properties and microstructure of E1, E2 and IC processed according to the manufacturer's recommendations are well documented in the literature. However, there are few studies evaluating the effects of microstructural anisotropy on the mechanical properties of dental ceramics. The aim of this work was to investigate the effects of microstructural anisotropy associated with the sample geometry on fracture behavior of three dental ceramics (two glass-ceramic prepared by heat-pressing method and a glass-infiltrated alumina composite).

EXPERIMENTAL

The dental ceramics used in this study are described in Table I. Materials were selected in order to provide varied microstructures: a) the glass-ceramic E1 is expected to have a homogeneous microstructure and isotropic properties; b) the glass-ceramic E2 is expected to have anisotropic properties due to alignment of lithium disilicate fibers; and c) the glass-infiltrated alumina composite IC, may have particle orientation, which could affect the mechanical properties.

Table I. Description of the materials used in this study.

Material	Manufacturer / Brand name	Manufacturer's description
E1	Ivoclar Vivadent / IPS Empress	Heat-pressed, leucite-based glass-ceramic, used for inlays, onlays, veneers and crowns
E2	Ivoclar Vivadent / IPS Empress 2	Heat-pressed, glass-ceramic with lithium disilicate, used as core material in crowns and bridges
IC	Vita Zahnfabrik / In-Ceram Alumina	Glass-infiltrated alumina composite, used as core material in crowns and bridges

Disks (12 mm in diameter and 2 mm in thickness) and bars (5 x 7 x 20 mm) of each material were produced according to each manufacturer instructions. Glass-ceramics were processed by the heat-press technique using a specific oven (EP 600. Ivoclar Vivadent, Schaan, Liechtenstein). In this method, a sprue (2 mm in diameter) was connected perpendicularly to the lateral (circumferential) surface of the disc or aligned parallel to the longest axis of the bar. These sprue configurations were similar to those used by Albakry et al.[11] The glass-ceramic ingots were heated to 1075°C (E1) or 920°C (E2) and then pressed to fill the sample cavity in a refractory mold. The In-Ceram Alumina composite was processed by infiltrating a lanthanum-silicate glass into a porous partially sintered alumina preform. Green bodies were prepared by slip casting an alumina particle slurry into a mold consisted of a gypsum substrate and lateral walls of silicone. The casting direction was perpendicular to the largest surface of the disc or bar. Green bodies were slightly sintered at 1120°C for 2 h in air to prepare preforms with sufficient strength to handle. A low sintering temperature was used to avoid undesirable shrinkage of alumina preforms. After applying the glass powder over the preform, glass was spontaneously infiltrated at 1110°C for 3 h in air. Both heating cycles, sintering and infiltration, were carried out in a specific furnace (InCeramat II, Vita Zahnfabrik, Bad Sackingen, Germany). All samples were machined following the guidelines in ASTM C 1161 and the tensile surface of each disc or bar was mirror polished using a polishing machine (Ecomet 3, Buehler, Lake Bluff, IL, USA) with diamond suspensions (45, 15, 6 and 1 μm). The final dimensions of the samples were: Ø12 x 1.1 mm for discs and 3 x 4 x ~20 mm for bars.

The fracture toughness, K_{Ic}, was determined by the indentation strength (IS) method for both types of sample geometry (disc and bar). By this method, a Vickers impression was introduced in the center of the polished surface and the fracture stress, σ_f, of indented specimen was measured in bending. The value of K_{Ic} was then calculated by:

$$K_{Ic} = \eta (E/H)^{1/8} (\sigma_f P^{1/3})^{3/4} \qquad (1)$$

where, η is a geometrical constant, which depends on crack and indenter shapes, E is Young's modulus, H is hardness and P is impression load.[19] The load P used for glass-ceramics E1 and E2 was 19.6 N and for the composite IC was 49.0 N. In the case of bar specimens, one diagonal of the Vickers indent was aligned parallel to the longer axis of bar. Immediately after indentation, the indented region was covered with a drop of silicone oil to minimize the effects of slow crack growth, and then subsequently broken in a bending test. Bar specimens where tested in a 3-point bending fixture with a span of 16 mm and disc specimens were fractured in a biaxial flexure test (piston-on-three-balls method), both using a universal testing machine (Syntech 5G, MTS, São Paulo, Brazil) at a crosshead speed of 0.5 mm/min. Vickers indentation and flexural tests for K_{Ic} measurement were carried out in laboratory air (~22°C, 60% RH). Five specimens were tested in each experimental condition.

The biaxial flexure strength, $\sigma_{f,b}$, of disc shaped samples was determined by piston-on-three-balls method with specimens immersed in artificial saliva (100 mL of KH_2PO_4 2.5 mM; 100 mL of Na_2HPO_4 2.4 mM; 100 mL of $KHCO_3$ 1.5 mM; 100 mL of NaCl 1.0 mM; 100 mL of $MgCl_2$ 0.15 mM; 100 mL of $CaCl_2$ 1.5 mM; and 6 mL of citric acid 0.002 mM; pH 7.0) at 37°C. The test was carried out at a stress rate of 1 MPa/s. The biaxial flexure strength, $\sigma_{f,b}$, was calculated by:[20]

$$\sigma_{f,b} = \frac{3P(1+v)}{4\pi t^2}\left[1 + 2\ln\frac{a}{b} + \frac{(1-v)}{(1+v)}\left\{1 - \frac{b^2}{2a^2}\right\}\frac{a^2}{R^2}\right] \qquad (2)$$

where P is the load at fracture, v is the Poisson's ratio, t is the thickness of the specimen, a is the radius of circle formed by three hardened steel balls equally spaced to support the specimen, b is the radius of the load carrying piston, and R is the radius of the specimen. Thirty specimens were tested for each material investigated. The strength data were analyzed using a two-parameter Weibull distribution, where the probability of failure P_f at or below a stress σ is given by:[21]

$$P_f = 1 - \exp\left[-\left(\frac{\sigma}{\sigma_0}\right)^m \right]$$ (3)

where m is the Weibull modulus and σ_0 is the characteristic strength (scale parameter). The Weibull parameters were estimated by maximum likelihood method (ASTM C 1239).

Elastic constants (E and v) were determined by ultrasonic pulse-echo method using a 200 MHz ultrasonic pulser-receiver (Panametrics 5900 PR, USA) and bulk density was determined by Archimedes' method.[22] Crystalline phases were identified by X-ray diffraction (XRD) analysis and the microstructural and fractographic analyses were performed using a scanning electron microscope (SEM) (JMS 6300, Jeol, Japan) coupled with an energy dispersive spectroscope (EDS) (Noram, USA).

Statistical analysis of K_{Ic} data was performed by means of one-way analysis of variance (ANOVA) and multiple comparisons were performed using Tukey's post-hoc test at a pre-set significance level of 5%. The results in the text are shown as average ± one standard deviation, except when indicated.

RESULTS
Fig. 1 shows SEM micrographs of the prepared samples.

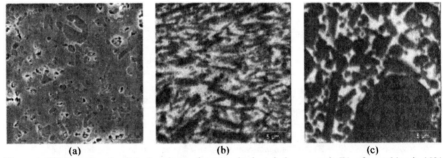

| (a) | (b) | (c) |

Figure 1. SEM images of polished surfaces of: (a) leucite-based glass ceramic E1, after etching in 10% HF solution; (b) lithium disilicate-based glass ceramic E2; and (c) glass-infiltrated alumina composite IC. Note: (a,c) – secondary electron image (SEI) and (b) – backscattered electron image (BEI).

Glass-ceramic E1 presented rounded tetragonal leucite particles dispersed homogeneously in the glassy matrix (the presence of these phases was confirmed by XRD analysis). The volume fraction of second-phase particles was 29% and the average leucite particle size was ~1 µm, but some long coalesced particles with length up to ~10 µm were also observed (Fig. 1a). No preferential particle orientation or alignment was observed throughout the microstructure of the E1 specimens. EDS analysis of leucite particles showed the presence of high amounts of Si, Al, K and O, as expected from stoichiometric leucite ($KAlSi_2O_6$), and indicated the presence of a small amount of Na in solid solution. The glassy matrix presented high amounts of Si, Al, Na, K and O and small amount of Ca, indicating that the non-crystalline phase is an alkaline aluminosilicate glass. Fig. 2a shows XRD patterns of glass-ceramic E1 after different treatments. No significant differences were observed among the different patterns, indicating that the heat-pressing did not cause significant crystallographic alignment of the leucite phase.

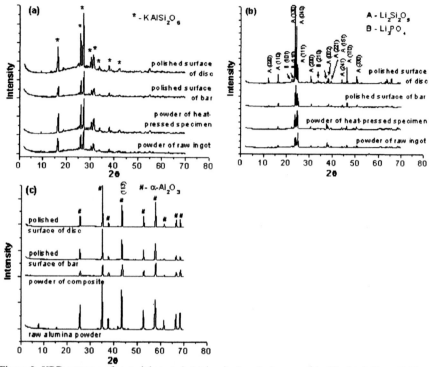

Figure 2. XRD patterns of materials tested: (a) leucite-based glass ceramic E1; (b) lithium disilicate-based glass ceramic E2; and (c) glass-infiltrated alumina composite IC.

The glass-ceramic E2 contained needle-like elongated particles with length of up to ~10 μm and thickness of up to ~1 μm (average aspect ratio of ~7 and volume fraction of 58%, Fig. 1b). XRD analysis showed the presence of lithium disilicate ($Li_2Si_2O_5$) as the main crystalline phase and also a small fraction of lithium phosphate (Li_3PO_4). The last phase could not be identified in SEM analysis and indicated that most of the particles observed in the micrograph of Fig. 1b are lithium disilicate. The needle-like particles tended to align parallel to their longer axis, at least at short distances, but the orientation of particles was not uniform. EDS analysis could not distinguish differences in chemical composition between needle-like particles and the glassy matrix, probably because of the small particle dimensions. Both consisted of mainly Si and O (Li could not be detected, since EDS analysis cannot detect low atomic number elements) and small amounts of K, Zn, Al and P. Fig. 2b shows XRD patterns of the glass-ceramic E2 after different treatments. Polished surfaces of both disc and bar shaped samples had similar XRD patterns, but with some variations in relation to the milled (powdered) heat-pressed E2 sample, mainly related to the relative peak intensities of the major $Li_2Si_2O_5$ phase, like the enhancement of diffraction peak of plane (130) and weakening of the peak associated with the plane (111). The main difference between the XRD patterns of disc and bar specimens was the higher relative peak intensity of the (170) in the disc sample (Fig. 2b). These results indicated that in both types of specimen geometries an alignment of crystalline planes of lithium disilicate phase occurred.

The glass-infiltrated alumina composite IC had a large particle size distribution with size ranging from ~1 to 20 μm (aspect ratio of ~5 and volume fraction of 65%, Fig. 1c). Alumina particles with different morphologies were observed: small equiaxed particles, large platelets, and faceted elongated particles (Fig. 1c). XRD indicated that α-Al_2O_3 was the only crystalline phase in the composite. EDS analysis indicated only Al and O in alumina particles and Al, Si, La and O, as main elements, with small amount of Ca, in the glassy matrix. Fig. 2c shows XRD patterns for the composite IC. Polished surfaces of both disc and bar shaped samples produced XRD patterns in general similar to that of the milled (powdered) composite. A slightly less intense peak for the (113) plane of α-Al_2O_3 in the surface of the disc was the only difference observed (Fig. 2c).

Table II shows the results of bulk density, ρ, Young's modulus, E, and Poisson's ratio, ν, of materials tested. All values of these properties increased significantly in the following order of materials: E1, E2 and IC. These behaviors seem to be related to the volume fraction and composition of the phases presenting in these materials.

Table II. Results of bulk density, ρ, Young's modulus, E, and Poisson's ratio, ν. For each property, values followed by the same superscript are statistically similar (p > 0.05).

Material	ρ (g/cm³)	E (GPa)	ν
E1	2.402 ± 0.005 [a]	66.1 ± 1.1 [a]	0.210 ± 0.007 [a]
E2	2.433 ± 0.008 [b]	99.3 ± 2.5 [b]	0.225 ± 0.012 [b]
IC	3.804 ± 0.023 [c]	276.0 ± 6.5 [c]	0.239 ± 0.008 [c]

The results of fracture toughness, K_{Ic}, of samples with disc and bar geometry (determined in inert condition) are shown in Table III. The K_{Ic} value of glass-ceramic E1 determined in disc sample was statistically similar to the value determined in bar sample, confirming that this material has an isotropic microstructure and mechanical properties. Therefore, heat-pressing did not produce any tendency toward crystallographic orientation in leucite-based glass-ceramics. The slightly higher mean value in bar sample (21%) may be related to the difference in loading mode (uniaxial versus biaxial). The other two materials, however, presented significantly higher K_{Ic} values when tested with the bar geometry. The significant differences of 55% and 33% for E2 and IC, respectively, can not be accounted only for the difference in loading mode. Therefore, heat pressing and sample geometry influenced the degree of anisotropy in the microstructure and mechanical properties of the lithium disilicate-based glass-ceramic E2, as previously reported.[11] For the glass-infiltrated alumina composite IC, the difference in K_{Ic} values obtained from disc and bar samples is also indicative of microstructure anisotropy. Comparing the K_{Ic} values among the materials tested, composite IC presented the highest value followed by glass-ceramic E2 and then by glass-ceramic E1, regardless of the sample geometry.

Weibull plots of biaxial flexural strength, $\sigma_{f,b}$, data of disc shaped samples tested in artificial saliva at 37°C are shown in Fig. 3 and the determined Weibull parameters are shown in Table IV. Considering the 95% confidence interval, the values of Weibull modulus, m, of the three tested materials were similar. A slightly higher mean m value, however, was observed in composite IC (m = 11.2), when compared to the values of glass-ceramics E1 and E2 (m ≈ 9.5). Significant differences were observed among the values of characteristic strength, σ_0 (Table IV). Composite IC had a significantly higher value of σ_0, when compared to the glass-ceramic E2, which had a significantly higher value than glass-ceramic E1. The same trend was observed for average value of biaxial flexural strength, $\sigma_{f,b}$ (Table III).

Table III. Results of fracture toughness, K_{Ic}, of samples with disc and bar geometry and biaxial flexural strength, $\sigma_{f,b}$, of disc shaped samples. Values followed by the same superscript are statistically similar (p > 0.05).

Material	K_{Ic} (MPa.m$^{1/2}$) disc	K_{Ic} (MPa.m$^{1/2}$) bar	$K_{Ic\ bar}$ / $K_{Ic\ disc}$	$\sigma_{f,b}$ (MPa) disc
E1	0.96 ± 0.03 [a]	1.16 ± 0.03 [a]	1.21	99 ± 12 [e]
E2	1.81 ± 0.18 [b]	2.81 ± 0.20 [c]	1.55	181 ± 20 [f]
IC	2.91 ± 0.20 [c]	3.87 ± 0.33 [d]	1.33	386 ± 32 [g]

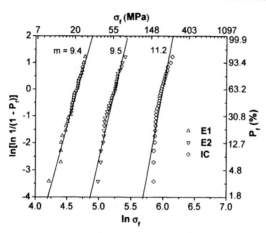

Figure 3. Weibull plot of biaxial flexural strength, $\sigma_{f,b}$, data of disc shaped samples tested in artificial saliva at 37°C.

Table IV – Strength distribution parameters: Weibull modulus, m, and characteristic strength, σ_0 (95% confidence interval is shown in parenthesis).

Weibull parameter	E1	E2	IC
m	9.4 (6.9 – 12.7)	9.5 (6.9 – 12.8)	11.2 (8.2 – 15.1)
σ_0 (MPa)	104.1 (99.6 – 108.6)	189.5 (181.3 – 197.7)	401.4 (386.8 – 416.1)

Typical macroscopic crack patterns of the specimens fractured in biaxial flexural test are shown in Fig. 4. Materials E1 and IC (Fig. 4a and 4c) presented usual crack patterns, where ramification of the crack usually results in more than two broken pieces, depending on the fracture stress. All specimens of glass-ceramic E2, however, fractured in two parts, regardless of the fracture stress. In fact, the fracture path was always curved (Fig. 4b). It seems like this anomalous crack pattern has not been described before for this material.

 (a) (b) (c)

Figure 4. Macroscopic crack patterns of the specimens fractured in biaxial flexural test of the material: (a) E1; (b) E2; and (c) IC.

Fig. 5 shows SEM images of fractured surfaces of materials tested. The fractured surface of glass-ceramic E1 was relatively smooth (Fig. 5a) compared to the other two materials. The rough fracture surfaces of glass-ceramic E2 (Fig. 5b) and composite IC (Fig. 5c) indicated that crack deflection around crystalline particles was the main toughening mechanism acting in these materials. Glass-ceramic E2 had many twist hackles on the fractured surfaces, which are indicative of successive changes in the plane of crack propagation, resulting in the formation of steps on the crack path. The fractured surfaces of composite IC presented elongated alumina particles, many aligned parallel to the planar surfaces of the discs (Fig. 5c). These particles are transversally fractured platelets, which are observed as large particles on the planar surface of the disc (Fig. 1c). Therefore, the composite IC presented anisotropic microstructure with alumina platelets tending to align parallel to the surface of the disc. The spatial distribution of alumina particles was defined during the slip casting of the alumina slurry, where the platelets tended to align their large surfaces perpendicularly to the gypsum mold (substrate).

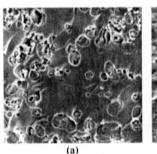 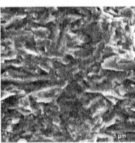 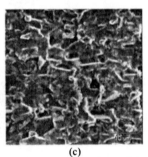
(a) (b) (c)

Figure 5. SEM images (SEI) of fractured surfaces of specimens broken in biaxial flexural test of the material: (a) E1; (b,c) E2; and (d) IC.

DISCUSSION

As expected, the leucite-based glass ceramic E1 had a homogeneous microstructure with rounded leucite particles, mostly isometric, and the specimen geometry (disc and bar) did not significantly affect the fracture toughness, K_{Ic} (Table 3). This means that the heat-pressing method alone did not induce the microstructural anisotropy. The results of lithium disilicate-based glass-ceramic E2 confirmed the previous report,[11] that K_{Ic} value is affected by specimen geometry, since the alignment of needle-like particles parallel to the long axis of bar specimen resulted in reinforcement of the specimen loaded in the transverse direction. The XRD patterns of polished surfaces of disc and bar were similar (Fig. 2b) and did not explain the difference observed on K_{Ic} values. As an additional experiment, XRD analyses of fractured surfaces of disc and bar specimens were conducted and the results are shown in Fig. 6. For disc specimen, XRD pattern was close to that observed on planar surface (Fig. 2b), but for the bar specimen, significant changes were observed, with strong enhancement of peak intensities of planes (111) and (002) of $Li_2Si_2O_5$ phase, showing strong crystallographic anisotropy in bar specimens.

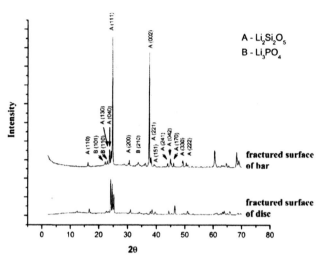

Figure 6. XRD patterns of fractured surfaces of the specimens with disc and bar geometries of lithium disilicate-based glass-ceramic E2.

Fig. 7 shows the relationship between biaxial flexural strength, $\sigma_{f,b}$, and fracture toughness, K_{Ic}, of the disc shaped samples. The dotted line in this figure represents $\sigma_{f,b} = 100.K_{Ic}$. Materials E1 and E2 followed this line, but the composite IC deviated from the line and had a higher $\sigma_{f,b}$ value. These results suggest that the toughening mechanism (crack deflection) in composite IC is more effective than in the glass-ceramic E2, and contributed to the higher Weibull modulus observed in this material (Table IV). The ratio of biaxial flexural strength, $\sigma_{f,b}$, values of E2 and E1 in this work was 1.83 (Table III). The value of this ratio was low, when compared to the literature. There are some studies that measured uniaxial flexural strength of both glass-ceramics (in air) and the observed flexural strength ratios (E2/E1) varied between 2.33 and 3.57.[7,8,10,23,24] These results indicated that the biaxial flexural strength of the glass-ceramic E2 with disc geometry was significantly lower than the uniaxial flexural strength with the bar geometry. In order to clarify if the needle-like particles had any type of alignment, a disc specimen with a polished surface was systematically indented with a Vickers diamond. It was observed that radial cracks that emanated from the Vickers indentations shifted from the diagonal lines of the impression, depending on the position of the impression on specimen. This shifting occurred because cracks propagated through the weakest paths and, therefore, indicated the particle alignment. Based on the radial crack shifting, a scheme of the particle distribution could be assessed, which is shown in Fig. 8. It can be seen in this figure that the long axis of the needle-like particles tended to align in the direction perpendicular to the direction of pressing. The proposed particle distribution explains the anomalous macroscopic fracture path observed in this material (Fig.

4b). Once a crack started to propagate, it would run preferentially through the weakest path, which is parallel to the long axis of the needle-like particles. Since these particles are aligned in a semi-circular manner (Fig. 8), the crack path follows this alignment. The twist hackles observed on the fractured surface corroborate this interpretation, since a continuous change of crack propagation plane is necessary to form the semi-circular crack path (Fig. 4b).

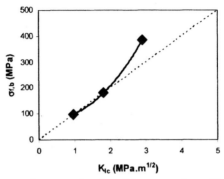

Figure 7. Biaxial flexural strength, $\sigma_{f,b}$, as a function of fracture toughness, K_{Ic}, of disc shaped samples. Dotted line represents $\sigma_{f,b} = 100.K_{Ic}$.

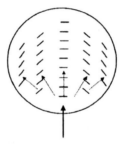

Figure 8. Schematic representation of the alignment of needle-like lithium disilicate particles in glass-ceramic E2 for disc shaped specimen. Larger arrow indicates the direction of pressing and small arrows indicate the possible pressing force directions that resulted in particle alignment.

In the case of composite IC, it is not clear yet why the K_{Ic} value of the disc shaped sample was lower than the value of the bar specimen. On the polished surfaces of the disc specimen of this material, however, some regions exhibited the tendency toward particle alignment, which may have caused the weakening of disc shaped sample. The lower K_{Ic} values of disc shaped samples for materials with anisotropic microstructure (E2 and IC) indicated that biaxial flexural testing of samples with disc geometry is preferred to evaluate the strength of dental materials, since the fracture tends to propagate by the weakest microstructural path. Usually a dental restoration has a complex geometry and is subjected to a multi-axial loading condition and, therefore, for safety reasons, the lowest strength and K_{Ic} values of the material are should be considered for design purpose and lifetime prediction analysis.

CONCLUSIONS

The results of this work showed that the mechanical properties (flexural strength and fracture toughness) of dental materials with non-isometric second-phase particles are affected by microstructural anisotropy caused by sample (or restoration) geometry and processing method. For the glass-ceramic reinforced with needle-like lithium disilicate particles (Empress 2), both sample geometry and the heat press method caused particle alignment. For the glass-infiltrated alumina composite (In-Ceram Alumina), the large alumina platelets tended to align their large surfaces perpendicularly to the direction of particle deposition during slip casting of green performs. The microstructural anisotropy enhances strength and fracture toughness when the material is stressed transversally to the particle alignment direction, but weakens it when fracture propagates parallel to this alignment. The biaxial flexural test of samples with disc geometry is recommended to evaluate the strength of material which tends to present microstructural anisotropy caused by the processing than uniaxial bending test of samples with bar geometry, since the fracture tends to propagate by the weakest microstructural path.

ACKNOWLEDGEMENTS

The authors acknowledge FAPESP (Fundação de Amparo à Pesquisa do Estado de São Paulo) and CNPq (Conselho Nacional de Desenvolvimento Científico e Tecnológico) for the financial support of the present research.

REFERENCES

[1] R. R. Braga, R. Y. Ballester, and M. Daronch, Influence of Time and Adhesive System on the Extrusion Shear Strength between Feldspathic Porcelain and Bovine Dentin, *Dent. Mater.*, **16**, 303-10 (2000).

[2] M. J.Cattell, R. L. Clarke, and E. J. Lynch, The Biaxial Flexural Strength and Reliability of Four Dental Ceramics-Part II, *J. Dent.*, **25**, 409-14 (1997).

[3] C. M. Gorman, W. E. McDevitt, and R. G. Hill, Comparison of Two Heat-Pressed All-Ceramic Dental Materials. *Dent. Mater.*, **16**, 389-95 (2000).

[4] W. Holand and G. Beall, Glass-Ceramic Technology, Westerville: The American Ceramic Society, (2002).

[5] J. K.Dong et al., Heat-Pressed Ceramics: Technology and Strength, *Int. J. Prosthodont*, **5**, 9-16 (1992).

[6] M. J. Cattell et al., The Biaxial Flexural Strength of Two Pressable Ceramic Systems, *J. Dent.*, **27**, 183-96 (1999).

[7] M. Guazzato et al., Strength, Fracture Toughness and Microstructure of a Selection of All-Ceramic Materials. Part I. Pressable and Alumina Glass-Infiltrated Ceramics, *Dent. Mater.*, **20**, 441-48 (2004).

[8] W. Holand et al., A Comparison of the Microstructure and Properties of the IPS Empress 2 and the IPS Empress Glass-Ceramics, *J. Biomed. Mater. Res.*, **53**, 297-303 (2000).

[9] J. L. Ong, D. W. Farley, and B. K. Norling, Quantification of Leucite Concentration Using X-ray Diffraction, *Dent. Mater.*, **16**, 20-25 (2000).

[10] M. Albakry, M. Guazzato, and M. V. Swain, Biaxial Flexural Strength, Elastic Moduli, and X-Ray Diffraction Characterization of Three Pressable All-Ceramic Materials, *J. Prosthet. Dent.*, **89**, 374-80 (2003).

[11] M. Albakry, M. Guazzato, and M. V. Swain, Fracture Toughness and Hardness Evaluation of Three Pressable All-Ceramic Dental Materials, *J. Dent.*, **31**, 181-88 (2003).

[12] S. C. Oh et al., Strength and Microstructure of IPS Empress 2 Glass-Ceramic After Different Treatments. *Int. J. Prosthodont*, **13**, 468-72 (2000).

[13]L. Probster and J. Diehl, Slip-Casting Alumina Ceramics for Crown and Bridge Restorations, *Quintessence Int.*, **23**, 25-31 (1992).

[14]J. R. Kelly, I. Nishimura, and S. D. Campbell, Ceramics in Dentistry: Historical Roots and Current Perspectives, *J. Prosthet. Dent.*, **75**, 18-32 (1996).

[15]S. O. Koutayas, All-Ceramic Posts and Cores: the State of the Art, *Quintessence Int.*, **30**, 383-92 (1999).

[16]J. W. Mclean, The Science and Art of Dental Ceramics, *Oper. Dent.*, **16**, 149-56 (1991).

[17]A. J. E. Qualtrough and V. Piddock, Ceramics Update, *J. Dent.*, 25, 91-95 (1997).

[18]M. Guazzato, M. Albakry, M.V. Swain, and J. Ironside, Mechanical Properties of In-Ceram Alumina and In-Ceram Zirconia, *Int. J. Prosthodont*, **15**, 339-46 (2002).

[19]P. Chantikul et al., A Critical Evaluation of Indentation Techniques for Measuring Fracture Toughness: II, Strength Method, *J. Am. Ceram. Soc.*, **64**, 539-43 (1981).

[20]D. K. Shetty et al., Biaxial Flexural Test for Ceramics, *Am. Ceram. Soc. Bull.*, **59**, 1193-97 (1980).

[21]D. Wu, J. Zhou, and Y. Li, Methods for Estimating Weibull Parameters for Brittle Materials, *J. Mater. Sci.*, 41, 5630-38 (2006).

[22]H. N. Yoshimura et al., Effect of Porosity on Mechanical Properties of a High Purity Alumina, Proceedings of the *107th Annual Meeting of The American Ceramic Society*, 2006.

[23]H. Fischer, G. Dautzenberg, and R. Marx, Nondestructive Estimation of the Strength of Dental Ceramic Materials, *Dent. Mater.*, **17**, 289-95 (2001).

[24]A. Della Bona, J. J. Mecholsky, Jr., and K. J. Anusavice, Fracture Behavior of Lithia Disilicate- and Leucite-Based Ceramics, Dent. Mater., **20**, 956-62 (2004).

Silicon Carbide, Carbon and Oxide Based Composites

MECHANICAL PROPERTIES OF Hi-NICALON S AND SA3 FIBER REINFORCED SiC/SiC MINICOMPOSITES

C. Sauder, A. Brusson, J. Lamon
University of Bordeaux / CNRS
Laboratoire des Composites Thermostructuraux
3, allée de la Boétie
33600 Pessac, France
lamon@lcts.u-bordeaux1.fr

ABSTRACT

The tensile behavior of CVI SiC/SiC reinforced with Hi-Nicalon S or SA3 fibers is investigated. Minicomposite test specimens were used. Minicomposites are reinforced by single tows. The mechanical behavior was correlated with microstructural features including tow failure strength and interface characteristics. The Hi-Nicalon S fiber reinforced minicomposites exhibited a conventional response, comparable to that one observed on composites reinforced with as-received Nicalon or Hi-Nicalon S fibers and possessing weak fiber/matrix interfaces. The SA3 fiber reinforced minicomposites exhibited stronger interfaces and premature failures. The behavior was improved after heat treatments.

INTRODUCTION

Next generations of nuclear reactors will require structural materials that retain excellent mechanical properties at high temperatures, in a hostile environment. The SiC/SiC composites made via Chemical Vapor Infiltration (CVI) reinforced with recently developed near stoichiometric fibers appear as promising candidates [1, 2].

Like CVD/CVI SiC, Hi-Nicalon S and SA3 fibers exhibit stability under irradiation, in terms of mechanical properties, structure and microstructure. The previous generations of SiC/SiC composites made via CVI and reinforced with non stoichiometric fibers (such as Nicalon or Hi-Nicalon SiC fibers), possess several interesting mechanical properties like high toughness, high strength and damage tolerance. Thus, it is necessary to investigate the mechanical behavior of CVI SiC/SiC composites reinforced with Hi-Nicalon S and SA3 fibers.

In this paper, primary emphasis was placed on the correlation of mechanical behavior with fiber/matrix interface resistance, since tailoring the interface is a way to control the mechanical behavior of composites [3-5]. Minicomposites are convenient specimens for interface evaluation. Minicomposites are unidirectional composites reinforced with single tows. The minicomposite approach to composite design and investigation has been addressed in several papers [6, 7]. Interface characteristics can be extracted from features of the tensile stress-strain curve [7, 8]. The interface shear stress is commensurate with reciprocal debond length. Thus it is considered that the interface shear stress provides a satisfactory estimate of fiber/matrix bond. This is a convenient parameter for comparing interface strengths: the higher the interface shear stress, the shorter the debond and the stronger the interface.

2- EXPERIMENTAL

2.1- Specimen preparation

SiC/SiC minicomposites were manufactured via Chemical Vapor Infiltration of Hi-

Nicalon S (Nippon Carbon, Japan) or Tyranno SA3 (Ube, Japan) tows [7] (Table 1). The tows were coated with either Pyrocarbon (150 nm thick) or a (PyC/SiC)$_5$ multilayer. The (PyC/SiC) multilayer interphase contained 5 layers of PyC alternating with SiC layers (30 nm thick each). A few minicomposites were heat treated, with a view to ordering microstructure. They are identified by prefix T (Table 2).

The cross sectional area of minicomposites was determined on SEM (Scanning Electronic Microscope) micrographs of polished cross sections, using image analysis. Errors may have been introduced because the surface of specimens may have experienced damage during polishing.

The cross sectional area of tows was determined from specific mass and density (table 1). This method provides an average value over entire specimen length. Fractions of fibers and matrix within minicomposites were determined from the respective cross sectional areas of tows and minicomposites (Table 2).

	Hi-Nicalon S	SA3
Density (g/cm^3)	2.98 (2.94)	3.1 (2.95)
Number of fibres	500	1600
Average diameter (μm)	13	7 (6.99)
Specific mass (g/1000m)	193 (191)	190 (192)
Young's Modulus (GPa)	372 (375)	387 (385)
Thermal Conductivity (W m^{-1}*K^{-1})	18	65

Table 1: Summary of fiber characteristics, provided by the vendor, or measured in-house (data between brackets).

Minicomposites		Fibres	Interphase	PyC layer (nm)	V$_m$(%)	Number of tests
HNS/PyC$_{100}$/SiC		Hi-NicalonS	PyC	150	53	10
THNS/PyC$_{100}$/SiC	treated					3
SA3/PyC$_{100}$/SiC		SA3	PyC	150	60	7
TSA3/PyC$_{100}$/SiC	treated					4
HNS/(PyC$_{30}$/SiC)$_5$/SiC		Hi-NicalonS	(PyC/SiC)$_5$	30*5	53	2
THNS/(PyC$_{30}$/SiC)$_5$/SiC	treated					3
SA3/(PyC$_{30}$/SiC)$_5$/SiC		SA3	(PyC/SiC)$_5$	30*5	59	2
TSA3/(PyC$_{30}$/SiC)$_5$/SiC	treated					2

Table 2: Characteristics of minicomposites and number of specimens tested.

2.2- Tensile tests

Uniaxial tension tests were performed at room temperature at a constant strain rate (50 μm/mn). The load was measured using a 500 N load cell. The minicomposite elongation was measured using two-parallel linear-variable differential transformer (LVDT) extensometers that were attached to the grips. Extensometers were located on each side of specimens, in order to control specimen alignment.

Minicomposite ends were affixed within metallic tubes using glue. The tubes were gripped into the testing machine. Gauge length (distance between inner ends of tubes) was 25 mm. The gripping system compliance (C$_S$) is needed in order to substract the gripping system deformation measured by the extensometer. C$_S$ was determined on dry tows with various gauge lengths (25, 50, 75 and 1000 mm): C_S = 0.3 µm/N with SA3 tows and 0.36 µm/N with Hi-Nicalon S tows.

Unloading-reloading cycles were conducted on a few specimens of each batch, in order to evaluate the interfacial shear stress. The interfacial shear stress was extracted from the width of hysteresis loop measured during the last unloading-reloading cycle, just before ultimate failure of minicomposite. The following equation which has been established elsewhere, was used [8]:

$$\tau = \frac{b_2 N(1 - a_1 V_f)^2 R_f}{2V_f^2 E_m} \left(\frac{\sigma_p^2}{\delta\Delta} \right) \left(\frac{\sigma}{\sigma_p} \right) \left(1 - \frac{\sigma}{\sigma_p} \right) \tag{1}$$

with

$$a_1 = \frac{E_f}{E_c} \tag{2}$$

$$b_2 = \frac{(1+v)E_m\left[E_f + (1-2v)E_c\right]}{E_f\left[(1+v)E_f + (1-v)E_c\right]} \tag{3}$$

where σ is the applied stress in the unloading sequence that corresponds to δΔ, σ_p the initial stress level at unloading, E_c the Young's modulus of the minicomposite, R_f the fiber radius, v the Poisson's ratio (v= v_m = v_f), E_m the Young's modulus of the matrix, E_f that of fiber, and V_f the volume fraction of fiber.

The number of matrix cracks in the gauge length (N) was determined by SEM inspection of specimens after failure. In order to detect all the cracks, the specimens were etched using Murakami's reagent, to get a selective attack of SiC at the surface of cracks.

The test on dry tows, also provided tension stress-strain curves from which various interesting data were extracted. These data are useful for analysis of minicomposites behavior in the frame of a microstructure/properties relationships approach. The fracture surface of specimens was examined by SEM after the tests.

3- RESULTS AND DISCUSSION

3.1- Material characterization

Figure 1 shows images of minicomposites. The thickness of matrix in the interior was around 2 µm, whereas the external layer was around 14 µm thick. The thickness of interphases was close to the desired values, indicated in table 2.

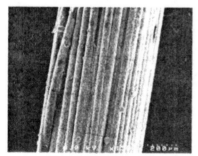

Figure 1: SEM images showing a minicomposite (Hi-NicalonS/(PyC30/SiC)5/SiC)

3.2- Tensile behavior of tows

Figure 2 shows the force-deformation curves obtained under tension (gauge length 25 mm). Table 3 summarizes the properties that were extracted from the stress strain behavior. The fiber statistical parameters and the fraction of individual fiber breaks were determined using equations available in [9-11]. The validity of statistical parameters was assessed by comparing experimental stress strain curves to predictions (figure 2). Equations are available in [11]. Note that both SA3 and HiNicalon S tows exhibit comparable properties. It is worth pointing that the strain-to-failure of tows is close to 0.7%. It is comparable to that one displayed by other SiC fibers (Hi-Nicalon and Nicalon).

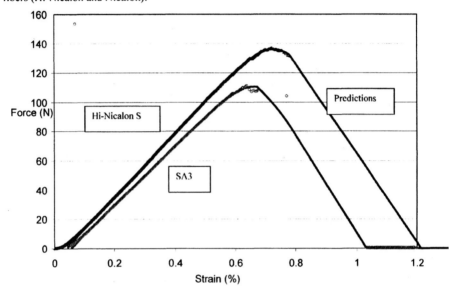

Figure 2: Tensile stress-strain behaviour of as received Hi-NicalonS and SA3 tows

Fibers		F_r N	E GPa	ε_r %	σ_l MPa	σ_{tow} MPa	γ %	α_c %	m	σ_o MPa
Hi- Nicalon S		120	319	0.73	1338	2477	14.1	12.3	6.4	1289
	Standard deviation	10	15	0.02	232	75	4.15	3.6	1.6	161
SA3		118	312	0.68	1231	2412	19.5	6.0	8.2	1353
	Standard deviation	5	3.5	0.04	111	122	0.91	0.97	2.1	155

Table 3: Main properties of tows (failure load F_r, Young's modulus E, strain-to-failure ε_r, proportional limit σ_l, strength σ_{tow}, fraction of fibers broken prior to loading γ, critical fraction of individual fiber breaks α_c) and single filaments (Weibull modulus m, scale factor σ_o) extracted from the tensile stress-strain curves.

3.3- Tensile behavior of minicomposites

Typical stress-strain curves obtained for the minicomposites are shown on figures 3 and 4.

The minicomposites reinforced with Hi-Nicalon S fibers exhibit conventional composite behavior that has been previously observed on SiC/SiC CVI minicomposites reinforced with Nicalon or Hi-Nicalon fibers [6, 7]. The force-deformation curves are markedly non linear, and they display the following typical features that reveal failures:

(a) an initial linear domain of elasticity,
(b) a curved region resulting from matrix cracking and associated fiber debonding. This region is bounded by a deformation < 0.25%, which indicates saturation of matrix cracking. This deformation can be considered as rather small.
(c) a linear domain after saturation, attributed to elastic deformation of fibers,
(d) a slight curvature preceding maximum force, indicating fiber breaks,
(e) ultimate failure occurred at deformations close to 0.7%, which compare fairly well with the strains-to-failure measured on tows.

The following interesting conclusions can be drawn from these features, on the basis of microstructure/properties relations that have been found in previous papers [6, 7, 12]:

- multiple matrix cracking was limited, crack density was small so that long debond cracks were probably induced by matrix cracks and fiber/matrix interfaces were rather weak.
- Ultimate failure was dictated by the failure of tows which indicates that potential contribution of fibers to mechanical behavior was not altered. The strength of fibers remain unchanged with respect to the reference one.

The tensile behavior displayed by the SA3 fiber reinforced minicomposites appears to be less complete than the previous one (figure 4): deformations are significantly smaller (< 0.20%, Table 4), so that the force-deformation curve displays only the initial elastic domain and a reduced non linear domain of matrix cracking (figure 4). This indicates that some of the previously mentioned phenomena did not occur (such as matrix cracking saturation) and that ultimate failure was not controlled by the tows.

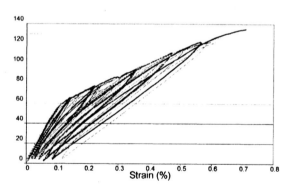

Strain (%)

Figure 3: Typical behaviour obtained with the Hi-NicalonS reinforced minicomposites (HNS/PyC$_{100}$/SiC mincomposite)

The treated specimens exhibit more complete stress-strain curves (figure 4), with deformations which tend to be as high as those that were measured on the tows. The tensile behavior is now controlled by the tows, which suggests that the premature failure of the minicomposites which have not been treated, cannot be attributed to fiber weakening during processing. The CVI process uses moderate temperatures ($\approx$ 1000°C), and data obtained on composites reinforced with Nicalon and Hi-Nicalon fibers show that fibers are not degraded under these conditions of temperature. Thus, it is more likely to deduce from these results, that fiber/matrix bonding was released during high temperature treatment. It is worth pointing out that similar effect has been obtained earlier on SiC/SiC CVI composites reinforced with Nicalon fibers [13]. The phenomena that lead to interface weakening have not been identified, at this stage. It can be conjectured that changes in the carbon layers nanostructure are involved.

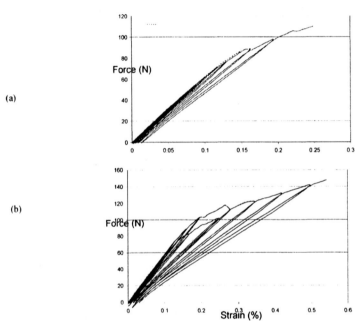

Figure 4: Typical behaviour obtained with the SA3 fiber reinforced minicomposites: (a) SA3/PyC100/SiC, (b) TSA3/PyC100/SiC minicomposites.

Table 4 reports the interface shear stresses that have been extracted from the hysteresis loops, as well as pull out lengths and matrix crack spacing distances. These latter data provide also a measure of interface strength. Both sets of data indicated the same trends, and they support the above statements. The interface shear stresses (τ) obtained for the minicomposites reinforced with Hi-Nicalon S fibers are quite low (Table 4). Weak interfaces could be logically expected for this fiber/interphase system. The current values of τ fall within the range of data determined on SiC/SiC minicomposites reinforced with Nicalon or Hi-Nicalon fibers [7, 12, 14]. Furthermore, it

can be noticed from table 4 that smaller τ were estimated on specimens that had been treated at high temperature. Much larger interface shear stresses were determined on the minicomposites reinforced with SA3 fibers, but τ was lowered tremendously by high temperature treatments, to values that can be compared to values obtained with Nicalon or Hi-Nicalon fibers [7, 12, 14]. Although they are high, the above τ values are associated to matrix cracking and debonding (table 4). Therefore, since they refer to a composite behavior, they cannot explain the premature failure of minicomposites. Instead, this premature failure must be attributed to:
- either a non uniform load distribution through specimen, that leads to high stresses locally,
- or fiber interactions that cause tow weakening [11]. These interactions result from local strong fiber/matrix bonds or fibers that are in contact which implies a wide distribution of interface strengths. The latter phenomenon is consistent with the effect of heat treatments on fiber/matrix bonds.

Higher τ were obtained on those minicomposites possessing multilayer interfaces and reinforced with Hi-Nicalon S fibers. Anyway, this trend must be considered with caution, since the number of tested specimens was not large. In a previous paper, the clear influence of multilayered interphases was not established due to scatter in interface shear stresses which appeared to be independent of PyC layer thickness [7].

An opposite trend can be noticed from the τ data measured on minicomposites reinforced with SA3 fibers. As pointed out above, a larger number of specimens would be required to assess this trend. But, it is interesting to compare crack patterns in single layer interphases and in multilayers. It can be seen from figure 5 that crack deviation occurred at the surface of fibers in the Hi-Nicalon S reinforced specimens, and that there are earlier deviations within the multilayers, which may cause τ increases [5, 7].

Minicomposites	F_r (N)	σ_r (MPa)	ε_r (%)	F_l (N)	σ_l (MPa)	τ (MPa)	l_p (μm)	d_s (μm)
HNS/PyC$_{100}$/SiC	119 (±5)	882 (±40)	0.61 (±0.06)	41 (±5)	304 (±27)	14.46 (-)	1000	300
THNS/PyC$_{100}$/SiC $^{(§)}$	66 (-)	487 (-)	0.17 (-)	27 (-)	200 (-)	5.43 (-)	8700	250
SA3/PyC$_{100}$/SiC	101 (±5)	614 (±40)	0.20 (±0.06)	64 (±5)	400 (±56)	223 (±8)	112	150
TSA3/PyC$_{100}$/SiC $^{(§)}$	126 (±22)	765 (±198)	0.38 (±0.15)	70 (±11)	500 (-)	104 (-)	345	150
HNS/(PyC$_{30}$/SiC)$_5$	134 (±3)	981 (±19)	0.65 (±0.02)	51 (±1)	370 (±20)	38 (-)	781	100
THNS/(PyC$_{30}$/SiC)$_5$$^{(§)}$	98 (-)	718 (-)	0.46 (-)	55 (-)	402 (-)	(*)	2844	(*)
SA3/(PyC$_{30}$/SiC)$_5$	101 (±20)	625 (±100)	0.15 (±0.01)	60 (±5)	400 (±40)	81 (-)	131	350
TSA3/(PyC$_{30}$/SiC)$_5$$^{(§)}$	96 (-)	607 (-)	0.24 (-)	52 (-)	405 (-)	(*)	60	(*)

(-) small sample size (*) not available (§) heat treated minicomposites

Table 4 : Main features of the tensile behaviour of minicomposites determined from the force-deformation curves (F_i is proportional limit) and from SEM fractography (fiber pull out length l_p, crack spacing distance at saturation d_s).

In the SA3 fiber reinforced minicomposites, crack deviation was also located at the surface of fibers when the interphase was a single layer. By contrast, it was detected within the multilayered interphase, whereas the SiC layer was found strongly bonded to the fiber (figure 5). This difference in crack deviation may explain the trend that was observed. This trend may be enhanced by load transfers through a larger surface when the first layer of SiC remains bonded to the fiber.

SA3/(PyC30/SiC)5/SiC

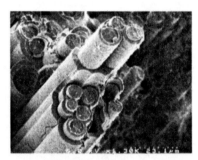

TSA3/(PyC30/SiC)5/SiC

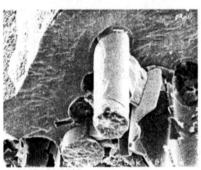

SA3/PyC100/SiC

HNS/(PyC$_{30}$/SiC)$_5$)SiC

Figure 5: Typical interface crack patterns detected in Hi-Nicalon S and SA3 fiber reinforced minicomposites.

4- CONCLUSIONS

The tensile behavior of minicomposites was correlated with microstructure features. Trends for Hi-Nicalon S fiber reinforced minicomposites are similar to those observed with earlier generations of SiC fibers (namely Nicalon and Hi-Nicalon): weak interfaces, limited matrix cracking and ultimate failure controlled by tows. A different trend was observed on the SA3 fiber reinforced minicomposites: stronger interfaces, premature failure attributed to non uniform fiber/matrix bonding, and ultimate failure is not controlled by tows. This behavior was changed after heat treatments which caused fiber/matrix bond release, leading to a composite response with ultimate failure controlled by tows.

In the minicomposites with multilayered interphases, opposite trends were observed on the Hi-Nicalon S and on the SA3 fiber reinforced minicomposites. The composite behavior was kept for the Hi-Nicalon S reinforced minicomposites, whereas it was degraded for the SA3 fiber reinforced minicomposites.

REFERENCES

[1]T. Muroga, M. Gasparotto and S.J. Zinkle, "Overview of materials research for fusion reactors", *Fusion Eng. Des.*, 61-62, 13-25 (2002).

[2]A. Hasegawa, A. Kohyama, R.H. Jones, L.L. Snead, B. Riccardi and P. Fenici, "Critical issues and current status of SiC/SiC composites for fusion", *J. Nuclear Materials*, 283-287, 128-137 (2000).

[3]R. Naslain, "Fiber-matrix interphases and interfaces in ceramic matrix composites processed by CVI", *Composite Interfaces*, 1 [3] 253-286 (1993).

[4]H.C. Cao, E. Bischoff, O. Sbaizero, M. Rühle, A.G. Evans, D.B. Marshall and J.J. Brennan, "Effect of interfaces on the properties of fiber-reinforced ceramic", *J. Am. Ceram. Soc.*, 73 [6] 1691-1699 (1990).

[5]C. Droillard, J. Lamon, "Fracture toughness of 2-D woven SiC/SiC composites with multilayered interphases", *J. Am. Ceram. Soc.*, 79 [4] 849-858 (1996).

[6]N. Lissart, J. Lamon, "Damage and failure in ceramic matrix minicomposites: experimental study and model", *Acta Mater.*, 45 [3] 1025 (1997).

[7]S. Bertrand, P. Forio, R. Pailler, J. Lamon, "Hi-Nicalon/SiC minicomposites with (Pyrocarbon/SiC)n nanoscale multilayered interphases", *J. Am. Ceram. Soc.*, 82 [9] 2465-73 (1999).

[8]J. Lamon, F. Rebillat, A.G. Evans, "Microcomposite test procedure for evaluating the interface properties of ceramic matrix composites", *J. Am. Ceram. Soc.*, 78 [2] 401-405 (1995).

[9]P. Forio, F. Lavaire, J. Lamon, "Delayed failure at intermediate temperatures (600°C – 700°C) in air in silicon carbide multifilament tows", *J. Am. Ceram. Soc.*, 85 [7] 888-893 (2004).

[10]European Standard ENV 1007-5 "Determination of distribution of tensile strength and of tensile strain to failure of filaments within a multifilament tow at ambient temperature" CEN TC 184, European Committee for Standardization, Brussels, Belgium, April 1998.

[11]V. Calard, J. Lamon, "Failure of fiber bundles", *Composites Science and Technology*, 64 (2004) 701-710.

[12]J. Lamon, "CVI SiC/SiC composites" chapter 3 in "Handbook of ceramics and glasses", Edited by Narottam P. Bansal, Kluwer Academic Publishers, New York, USA, pp. 55-76 (2005).

[13]L. Guillaumat, J. Lamon, "Mechanical behaviour and damage of ceramic matrix composites" (in French), Revue des Composites et Matériaux Avancés, 9 [2], 183-203 (1999).

[14]S. Bertrand, R. Pailler, J. Lamon "Influence of strong fiber/coating interfaces on the mechanical behaviour and lifetime of Hi-Nicalon/(PyC/SiC)n/SiC minicomposites", *J. Amer. Ceram. Soc*, 84 [4] 787-94 (2001).

THE EFFECT OF HOLES ON THE RESIDUAL STRENGTH OF SIC/SIC CERAMIC COMPOSITES

Ojard, G.[2], Gowayed, Y.[3], Santhosh, U.[4], Ahmad J.[4], Miller, R.[2], and John, R.[1]

[1] Air Force Research Laboratory, AFRL/MLLMN, Wright-Patterson AFB, OH
[2] Pratt & Whitney, East Hartford, CT
[3] Auburn University, Auburn, AL
[4] Research Applications, Inc., San Diego, CA

ABSTRACT

Continued interest in ceramic matrix composites for the use in high temperature applications necessitates a comprehensive understanding of the effect of time-dependent loading in various environments on the material's mechanical response. This understanding has to include the effect of structural features, such as holes, on the material performance. In this work, residual strength tests were conducted for samples with 2.286, and 4.572 mm average diameter holes, located at the center of the specimen, after time-dependent experiments conducted in air. Samples were subjected to creep and dwell fatigue tests at 1204 °C under net-section stresses ranging from 55.16 to 165.48 MPa for durations ranging from 10 to 2400 hours prior to residual strength experiments at room temperature. Data acquired from residual strength testing for samples with holes are analyzed and compared to similar data from standard samples.

INTRODUCTION

Testing is conducted to understand the long-term behavior of Ceramic Matrix Composites (CMCs) under conditions of sustained load at high temperatures. Examples of long-term environments would include ground base turbines for power generation where CMCs are being considered for combustor liners, turbine vanes and shroud applications. These applications can see design times of up to 30,000 hours. Such long term applications are working to leverage the high temperature material capability while taking advantage of reduced cooling and durability improvements that CMCs can provide over typical metals without cooling air that is needed for the nickel base superalloys as a possible replacement. This paper reports data and analysis of specimens with holes subjected to time-dependent tests followed by a residual strength experiments conducted at room temperature and compared to data on standard specimens published in previous years.

EXPERIMENTAL PROGRAM

Materials and Manufacturing

The material chosen for the study was the Melt Infiltrated SiC/SiC CMC system, which was initially developed under the Enabling Propulsion Materials Program (EPM) and is still under further refinement at NASA-Glenn Research Center (GRC). This material system has been systematically studied at various development periods and the most promising was the 01/01 Melt Infiltrated iBN SiC/SiC (01/01 is indicative of the month and year that development was frozen). There is a wide set of data from NASA for this system as well as a broad historic database from the material development. This allowed a testing system to be put into place to look for key development properties which would be needed from a modeling effort and would hence leverage existing data generated by NASA-GRC.

The Sylramic® fiber was fabricated by DuPont as a 10 µm diameter stochiometric SiC fiber and bundled into tows of 800 fibers each. The sizing applied was polyvinyl alcohol (PVA). For this study, the four lots of fibers, which were used, were wound on 19 different spools. The tow spools were then woven into a 5HS balanced weave at 20 EPI. An in-situ Boron Nitride (iBN) treatment was performed on the weave (at NASA-GRC), which created a fine layer of BN on every fiber. The fabric was then laid in graphite tooling to correspond to the final part design (flat plates for this experimental program). All the panels were manufactured from a symmetric cross ply laminate using a total of 8 plies. The graphite tooling has holes to allow the CVI deposition to occur. At this stage, another BN coat layer was applied. This BN coating was doped with Si to provide better environmental protection of the interface. This was followed by SiC vapor deposition around the tows. Typically, densification is done to about 30% open porosity. SiC particulates are then slurry cast into the material followed by melt infiltration of a Si alloy to arrive at a nearly full density material. The material at this time has less than 2% open porosity.

After fabrication, all the panels were interrogated by pulse echo ultrasound (10 MHz) and film X-ray. There was no indication of any delamination and no gross porosity regions were noted in the panels. In addition, each panel had two tensile bars extracted for witness testing at room temperature. All samples tested failed above a 0.3% strain to failure requirement. Hence, all panels were accepted into the testing effort. Samples were cut out of the accepted plates and holes where drilled in diameters of 2.286 and 4.572 mm at the center of the specimens forming 20% of the width of the specimen.

TESTING PROCEDURE
Creep testing at 1204 °C (exposure)
    Creep testing is a test to determine the strain-time dependence of a material under a constant load. This test is also used to determine the long-term behavior of the material under combination of load and temperature. For ceramic matrix composites, when this testing is done in air, there is an added complexity of environmental exposure. A total of 12 creep tests were done on straight-sided tensile bars with two different central hole sizes (2.286 and 4.572 mm) at 1204 °C.

Dwell Fatigue at 1204 °C (exposure)
    Dwell fatigue testing, sometimes referred to as "low cycle fatigue tests", is the superposition of a stress hold at the peak stress of a fatigue cycle. Dwell fatigue tests were conducted for a dwell time of 2 hours with a 1 minute load and unload at the beginning and end of each cycle. Eighteen straight-sided tensile bars with two different central hole sizes (2.286 and 4.572 mm) were tested under 55.16, 110.32, and 165.48 MPa net-section stress levels at an R ratio of 0.05 at 1204 °C. Electronic data acquired were large in size and include load-reload tests at different time intervals. A Matlab code was developed to extract this data and allow characterization of the change in strain with time.

Residual strength tests at room temperature (post exposure)
    Samples subjected to time dependent testing in creep or dwell fatigue experiments (exposure) were subjected to cyclic tensile loading at room temperature until failure (post exposure). Elastic and creep strains from exposure experiments as well as strain to failure during post-exposure

experiments were recorded along with the slope of the stress strain curves for apparent modulus calculations.

RESULTS AND DISCUSSION

Strain, stress to failure and modulus values acquired for the 30 specimens with holes are listed in Table 1 along with time-dependent strain and exposure time that each specimen witnessed during exposure testing. The time under exposure varied from 10 hours to 2400 hours. Sample of stress strain diagrams is shown in Figure 1 and 2.

Table 1: Experimental data for samples tested in residual strength after dwell fatigue and creep experiments at different net section stress levels:

Dwell Fatigue or Creep	Net Section Stress (MPa)	Time (hours)	Dwell Fatigue Elastic Strain (loading) (mm/mm)	Dwell Fatigue Max Strain During Exposure (mm/mm)	Creep Strain During Exposure (mm/mm)	Residual Apparent Modulus (13-56 MPa) (GPa)	Residual UTS Net Section (MPa)	Residual Strain to Failure (mm/mm)	Apparent Re-Load Modulus (13-56 MPa) (GPa)
2.286 mm hole									
Dwell	55.16	10	0.000189	0.000305		313.7	349.58	0.0018	276.9
Dwell	55.16	100	0.000206	0.000427		298.3	420.60	0.0027	275.5
Dwell	55.16	10	0.000205	0.000298		309.3	379.23	0.0023	276.6
Dwell	55.16	100	0.000216	0.00038		309.8	392.33	0.0025	270.8
Dwell	165.48	500	0.000664	0.001648		292.5	336.92	0.0015	287.9
Dwell	165.48	524	0.000588	0.001259		302.2	325.85	0.0013	293.9
Dwell	165.48	436	0.000583	0.001665		288.2	302.40	0.0012	285.8
Dwell	110.32	100	0.000391	0.00134		301.4	341.30	0.0017	289.2
Dwell	110.32	10	0.000328	0.000519		301.6	372.33	0.0020	289.5
Dwell	110.32	100	0.000341	0.000591		314.4	404.74	0.0023	296.5
Creep	110.32	1004			0.0026	284.0	357.85	0.0023	262.9
Creep	55.16	2150			0.00293	261.9	264.21	0.0018	na
Creep	55.16	1129			0.00193	270.7	312.05	0.0022	228.6
Creep	110.32	1061			0.00163	270.5	368.19	na	245.1
Creep	55.16	2400			0.00387	259.7	309.90	0.0023	221.3
Creep	55.16	998			0.00078	279.0	325.95	0.0023	231.9
4.572 mm hole									
Dwell	110.32	100	0.00041	0.000824		289.2	335.10	0.0015	288.8
Dwell	110.32	10	0.000412	0.000546		303.9	330.96	0.0016	297.6
Dwell	110.32	100	0.000416	0.000637		303.5	343.37	0.0018	284.9
Dwell	55.16	10	0.000168	0.000125		304.7	346.82	0.0019	286.7
Dwell	55.16	100	0.000181	0.000351		293.0	341.30	0.0019	269.7
Dwell	110.32	10	0.000355	0.0007		300.4	330.96	0.0016	293.4
Dwell	55.16	100	0.000208	0.000364		316.4	363.19	0.0021	289.5
Dwell	55.16	10	0.000244	0.000293		316.5	352.33	0.0019	296.4
Creep	55.16	858			0.00373	286.3	314.41	0.0018	263.3
Creep	110.32	1041			0.00332	306.7	312.33	0.0015	287.6
Creep	55.16	1052			0.00113	297.0	282.14	0.0015	274.6
Creep	110.32	1046			0.00402	323.2	312.65	0.0014	306.1
Creep	110.32	1863			0.00515	279.5	210.61	0.0011	na
Creep	55.16	2065			0.00206	289.1	325.77	0.00183	261.5

na = not available

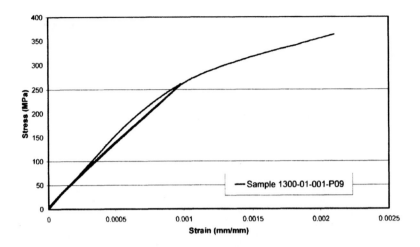

Figure 1. Residual strength for sample with 4.572 mm hole tested in dwell fatigue for 100 hours at 55.16 MPa

The effect of exposure time on residual strength is shown in Figure 3 and on the strain to failure is illustrated in Figure 4 along with data from standard samples as reference. Figure 5 shows the change of the value of strain to failure with the exposure strain for samples with and without holes.

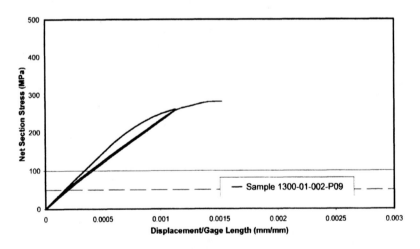

Figure 2. Residual strength for sample with 4.572 mm hole tested in dwell fatigue for 1052 hours at 55.16 MPa

It is important to note that the existence of the hole in the area of the gage length added an un-remediated artifact to the strain calculation and forced the use of an alternative term to strain, which is displacement/gage length. This limits the comparison between standard samples and samples with holes to be qualitative. Using this term, the effect of exposure time on residual strength is shown in Figure 3 and on the displacement/gage to failure is illustrated in Figure 4 along with data from standard samples as reference. Figure 5 shows the change of the value of strain to failure with the exposure strain for samples with and without holes. It can be seen from Figures 3 and 4 that there is a general reduction in residual strength and residual displacement/gage length with the increase in exposure time and the value of displacement/gage length accumulated during exposure. Standard specimens showed higher residual strength and residual displacement/gage length than specimens with holes. The impact of hole-diameter was not obvious and data for specimens with 2.286 mm holes are at the same level as those from specimens with 4.572 mm holes.

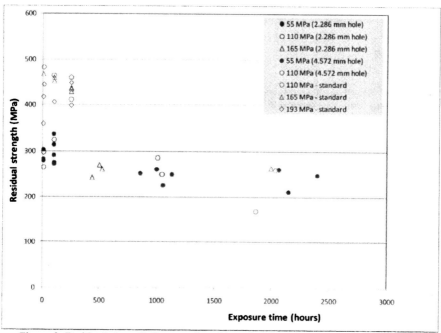

Figure 3. Residual strength for samples with and without holes vs. exposure time for samples subjected to various stress levels during initial testing

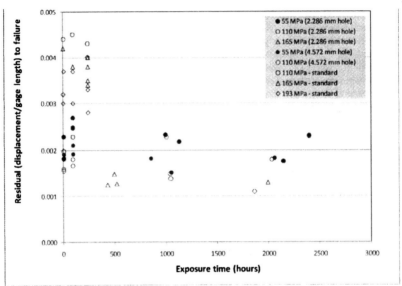

Figure 4. Residual (displacement/gage length) for samples with and without holes vs. exposure time for samples subjected to various stress levels during initial testing

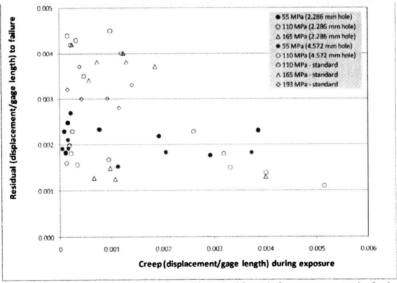

Figure 5. Residual strain for samples with and without holes vs. creep strain during exposure for samples subjected to various stress levels during initial testing

The depreciation of materials capabilities with time-dependent loading under dwell fatigue experiments, as well as fatigue experiments, can be attributed to the reduction in the ability of the fiber/coat and the coat/matrix material to carry shear strength. This concept was studied by Reynaud et al. [1,2], and used to characterize the micro-structural origin of the hysteresis by linking it to fiber/coat and matrix/coat interfacial shear assuming that under repeated loading the composite forms a series of interfacial damaged and undamaged zones confined between matrix cracks. Progressive wear at the interfaces between the fiber/coat and the coat/matrix is considered responsible for depreciation of the ability to carry interfacial stresses that starts in limited zones and progresses to cover most of the composite. The approach, as defined by Reynaud [1,2], can be illustrated by two scenarios as shown in Figure 6:

1. Increase in the area of hysteresis loops with the number of cycles leading to a decrease in ability to carry interfacial shear stress (τ) resulting from a local sliding at the fiber/coat or coat/matrix interface.
2. Decrease in the area of hysteresis loops with number of cycles leading to a decrease in the ability to carry interfacial shear stress resulting from a global sliding at all the interfaces.

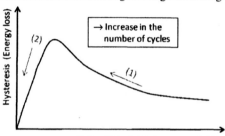

Figure 6: Decrease in interfacial shear stress with mechanical hysteresis following scenario 1 (right) and scenario 2 (left) [1]

As seen from Figure 7, the area of the hysteresis loops decreases with the number of loading cycles following the second scenario indicating a possible decrease in the ability to carry of interfacial shear stress with repeated loading. This can be calculated as follows [1]:

$$\frac{\Delta W}{W_e}(\tau) = \alpha\left(\frac{\tau}{\tau^*}\right)\left(\frac{1 - \frac{2}{3}\frac{\tau}{\tau^*}}{1 + \alpha\left(1 - \frac{\tau}{2 \cdot \tau^*}\right)}\right)$$

Where,

$$W_e = \frac{S}{2 \cdot E_x} \quad ; \quad \alpha = \frac{E_m \cdot V_m}{E_f \cdot V_f} \quad ; \quad \tau^* = \frac{\alpha r E_f S}{2 dE_x}$$

S = maximum stress applied, and r = radius of fibers, ΔW = area of hysteresis loop, d = mean distance between two neighboring cracks, E_f, V_f, E_m and V_m are the Young's moduli and volume fractions of fibers and matrix, respectively, and E_x is the Young's modulus of the composite in the load direction.

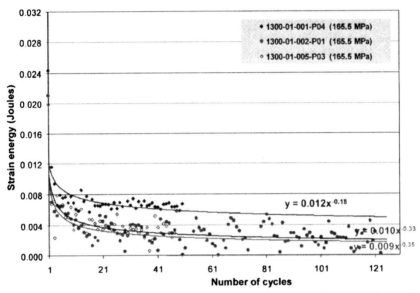

Figure 7. The effect of cycles during dwell fatigue experiments on the area of hysteresis loops of the stress-strain curves

CONCLUSIONS

Residual strength and displacement/gage length to failure of specimens with holes was presented in this work. It was observed that although the effect of the existence of the hole has an impact on the material response when compared to specimens without holes, the effect of the change in the hole size between 2.286 and 4.572 mm did not show a difference in response. A possible damage mechanism due to repeated loading evident in the reduction in the area of hysteresis loops was discussed.

ACKNOWLEDGMENTS

The Materials & Manufacturing Directorate, Air Force Research Laboratory under contract F33615-03-2-5200 and contract F33615-01-C-5234 sponsored this work

REFERENCES

1. Reynaud, P., Rouby, D. and Fantozzi, G., "Effects of temperature and of oxidation on the interfacial shear stress between fibres and matrix in ceramic-matrix composites", Acta mater., vol 46, No. 7, pp 2461-2469, 1998
2. Reynaud, P., "Cyclic fatigue of ceramic-matrix composites at ambient and elevated temperatures", Composites Science and Technology, vol. 56, pp. 809-814, 1996.

THROUGH THICKNESS MODULUS (E33) OF CERAMIC MATRIX COMPOSITES: MECHANICAL TEST METHOD CONFIRMATION

Ojard, G.[2], Barnett. T.[3], Calomino, A.[4], Gowayed, Y.[5], Santhosh, U.[6], Ahmaad, J.[6], Miller, R.[2], and John, R.[1]

[2]Pratt & Whitney, East Hartford, CT

[3]Southern Research Institute, Birmingham, AL

[4]NASA-Glenn Research Center, Cleveland, OH

[5]Auburn University, Auburn, AL

[6]Research Applications, Inc., San Diego, CA

[1]Air Force Research Lab, Wright-Patterson Air Force Base, OH

ABSTRACT

Considering the interest in and challenges of using ceramic matrix composites, a full understanding of the physical properties is needed to continue to move the technology forward. A property of interest to the design process is the through thickness modulus of the composite. Most high performance CMC material systems are produced in a relatively thin state that hinders generation of test specimens that can easily be instrumented and tested in the through-thickness direction. Past work by the authors has shown that a stacked disk method could measure the through thickness modulus at various temperatures with sufficient accuracy. Due to the nature of the stacked disk, several testing issues were present and a model material system was chosen to alleviate testing concerns. A monolithic ceramic system was used to verify the test method by using the material in its monolithic state deploying the same stacked disk method. Results and conclusions of this verification effort will be discussed.

INTRODUCTION

Ceramic matrix composites are considered for advanced engineering applications such as aggressive gas turbine engine applications and thermal protection systems due to the fact that the mechanical properties are relatively constant with temperatures up to the maximum use temperature (1). This is especially the case for turbine and combustor applications where there are high temperatures and possibly high stresses (2,3). These potential applications look to ceramic matrix composites due to the high temperature capability promised in strength at temperature with reduced or no cooling air requirements coupled with a lower density material. This can be seen in the wide breadth of work considering ceramic matrix composites in multiple applications (4).

As designs progress and become more challenging, there is a need to fully understand the physical properties of the material. A specific property of interest is the through thickness elastic modulus (E_{33}). This property is based on interactions between the fiber, fiber interface coating and the matrix as well as porosity as the material is pulled or compressed in the through thickness direction (z direction or 33 direction). The authors have documented a test method that can measure this property as a function of temperature (5). Since that time, a model material was used to review the test method arrive at, which relied upon testing a stack of discs greater than 25 mm in height. This was done to assure that the method is robust.

PROCEDURE

Material – CMC

Melt Infiltrated In-Situ BN SiC/SiC composite (MI SiC/SiC) is the material that was documented in the past work (5). The MI SiC/SiC system has a stochiometric SiC (Sylramic™) fiber in a multiphase matrix of SiC deposited by chemical vapor deposition followed by slurry casting of SiC particulates with a final melt infiltration of Si. The specific MI SiC/SiC tested had 36% volume fraction fibers using a 5 HS weave at 20 EPI with a cross ply lay-up. A cross section of the material is shown in Figure 1.

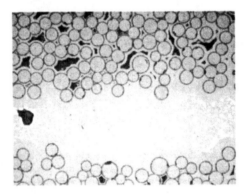

Figure 1. Cross sections of Melt Infiltrated In-Situ BN SiC/SiC Composite

Material – Monolithic (Test Material)

The material chosen as a test material is MACOR. MACOR is a machinable glass ceramic and is a fluorine rich glass with a composition approaching trisilicic fluorphlogopite mica ($KMg_3AlSi_3O_{10}F_2$) (6). The material was acquired in a rod form.

Method

The past method relied upon a stacked disk method (5). Enough disks were machined for one sample to allow an overall stack height greater than 27 mm. By having a stack greater than 25 mm, flags could be placed onto the stack to allow extensometry to be used over an 18-20 mm gage section. This would allow strain to be measured during compression.

For the MACOR material. the material was machined into disks as was the case for the CMC as well as leaving a solid rod of material greater than 25 mm in height for testing in compression as a comparison (where the interfaces from stacking would not be present).

All disks machined for this effort (CMC and MACOR) were machined with a center hole so that a centering rod of graphite could be used. The rod was machined shorter than the disk height so that it would not interfere or influence the measurement. This is shown in Figure 2.

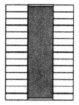

Figure 2. Schematic of stacked disks (method) with centering hole (rod)

RESULTS

The work done on the MI SiC/SiC system in the past paper is summarized in Figure 3 showing the modulus data as a function of temperature. The ultrasonic value (at room temperature) falls within the two experiments done on the CMC.

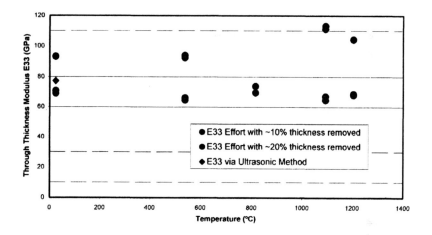

Figure 3. Through Thickness Modulus values versus temperature for MI SiC/SiC

The stress-strain for the CMC testing is shown in Figure 4. This figure shows that removing the surface asperities removes the initial compliance from the test. This also results in a higher modulus value since there is less porosity affecting the results.

The results from the MACOR rod and stacked disks are shown in Figure 5. As expected, the sold rod does not have the compliance issue seen in the stacked effort. The modulus value determined from the solid rod was found to be 66.6 GPa and the modulus value from stacked disk method was found to be 66.0 GPa. The literature value is published as 66.9 GPa (6).

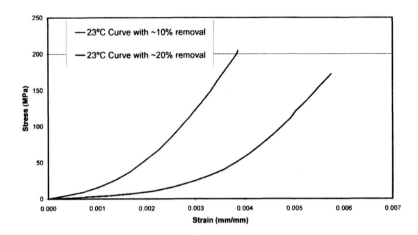

Figure 4. Stress-strain data at 24°C for modulus determination shown in Figure 3

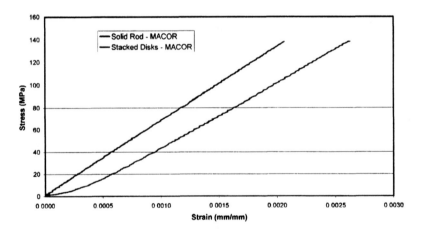

Figure 5. Stress-strain data at 24°C for the MACOR material (solid and stacked disks)

DISCUSSION

The testing done on the MACOR material shows that the use of a stacked disk does not affect the resulting determination of modulus. It does show that there is concern about taking into account the initial compliance of the stack. For the stacked MACOR experiment, it is clear that the stress-strain data up to 40 MPa needs to be excluded in any analysis of the modulus from this experiment. The testing of the CMC shows that an even higher stress level is required before looking into what the modulus value is. This was set as being 160 MPa for the CMC based on past work (5). Figure 4 clearly shows that the slope of the curves is leveling off at this stress level.

The MACOR differs from the CMC in that when it was machined into disks and then ground, there is not as much surface asperity to remove. This is clearly not the case for the CMC system studied. The surface is not uniform initially due to the 5 HS weave used in fabricating the ply. Once the plys are stacked (in this case all material was 8 ply thick), several CVI runs are done applying more material to the surface. As can be seen in Figure 4, as more material is removed, the initial compliance of the stack is reduced. This is due to removing the surface asperities. A balance must be struck between how much material is removed before the material is no longer representative of the CMC.

CONCLUSIONS

The method proposed of stacking disks of CMC material to determine the through thickness modulus of the material has been verified. It was found that stacking a known material to determine modulus and comparing that to a solid rod of material (for modulus) showed excellent agreement. Testing on a known material confirmed the need to look at the data outside of the initial compliance seen in the stress-strain data. Additional work should be taken on CMC systems to quantify the scatter seen in the method by looking at multiple samples within a given material system.

ACKNOWLEDGMENTS

The Materials & Manufacturing Directorate, Air Force Research Laboratory under contract F33615-01-C-5234 and contract F33615-03-D-2354-D04 sponsored portions of this work

REFERENCES

[1] Wedell, James K. and Ahluwalia, K.S., "Development of CVI SiC/SiC CFCCs for Industrial Applications" 39th International SAMPE Symposium April 11- 14, 1994, Anaheim California Volume 2 pg. 2326.
[2] Brewer, D., Ojard, G. and Gibler, M., "Ceramic Matrix Composite Combustor Liner Rig Test", ASME Turbo Expo 2000, Munich, Germany, May 8-11, 2000, ASME Paper 2000-GT-0670.
[3] Calomino, A., and Verrilli, M., "Ceramic Matrix Composite Vane Sub-element Fabrication", ASME Turbo Expo 2004, Vienna, Austria, June 14-17, 2004, ASME Paper 2004-53974.
[4] Chawla, K.K., Ceramic Matrix Composites, Kluwer Academic Publishers, Boston.
[5] Ojard, G., Barnett. T., Calomino, A., Gowayed, Y., Santhosh, U., Ahmaad, J., Miller, R., and John, R. "Through Thickness Modulus (E33) of Ceramic Matrix Composites: Mechanical Test Method Development", Ceramic Engineering and Science Proceedings. Vol. 27, no. 2, pp. 331-340. 2007
[6] http://www.accuratus.com/macorfab.html

THE EFFECTS OF Si CONTENT AND SiC POLYTYPE ON THE MICROSTRUCTURE
AND PROPERTIES OF RBSC

A. L. Marshall, P. Chhillar, P. Karandikar, A. McCormick, M. K. Aghajanian
M Cubed Technologies, Inc.
1 Tralee Industrial Park
Newark, DE 19711

ABSTRACT
        Composites of silicon carbide (SiC) and silicon (Si) are fabricated by the reactive
infiltration of molten Si into preforms of SiC particles and carbon. This product is often referred
to as reaction bonded silicon carbide (RBSC). SiC materials are used in many applications due to
their favorable properties including high hardness, high thermal conductivity, low thermal
expansion and high stiffness. Applications for such materials are numerous and can include
armor products, wear products and thermal management devices. This study attempts to further
optimize or tailor the physical, thermal and mechanical properties of RBSC by controlling the
silicon content and the SiC polytypes. The process variables studied in the present work include
the amount of carbon available for reaction and the type of SiC starting material. Powders with
predominant polytypes 6H-SiC and 3C-SiC were selected as the starting raw materials.
Combinations of the 6H, 3C, 2H, 4H, and 15R polytypes of SiC were present in the finished
RBSC ceramics. This study correlates the microstructure and phase composition with the
physical, thermal, and mechanical properties of the resultant ceramics. Increases in Young's
modulus and thermal conductivity are evident from increased SiC content. Most of the
conventional RBSC products are formed with 6H-SiC as a starting powder. Successful
fabrication of RBSC was proven possible with 3C-SiC as a starting powder. The 3C polytype
remained unchanged through the reaction bonding process. Results show clear effects of SiC
polytype and Si:SiC ratio on the thermal properties of the final ceramic. Effects of SiC polytype
on mechanical properties were small.

INTRODUCTION
        Reaction bonded SiC (RBSC) ceramics are most commonly manufactured by the
infiltration of molten Si into preforms of SiC particles plus carbon. During the infiltration
process, the Si and carbon react to form SiC. However, it is not possible to fully react all Si to
SiC without "canning-off" the infiltration process. Thus, an RBSC ceramic is generally a Si/SiC
composite consisting of predominantly SiC and some residual Si phase [1-3]. RBSC ceramics
are used in a wide range of applications from armor and wear components to thermal
management and even structural elements. This range of applications stems from the collectively
useful properties these composites exhibit such as high hardness, high stiffness, high thermal
conductivity, low density and low thermal expansion. Understanding how the properties change
with the predominant SiC polytypes is important as it relates to the development of optimized
ceramic compositions.
        Silicon carbide is known to crystallize in several crystallographic modifications, all
having the same "a" parameters (3.078 Å) but different "c" parameters, known as polytypes.
The structural reason for this phenomenon is a low stacking fault energy and the possibility to
form different modes of stacking of two-dimensional, structural, compatible units along a
definite direction. Some 200 polytypes of SiC are known to exist, but the most common are 3C,

2H, 4H, 6H and 15R. The digits stand for the repeatability of the stacking layer and the alphabet indicates the cubic, hexagonal and rhombohedral structure of these polytypes [4]. The 3C polytype is often referred to as beta-SiC and the 6H, 2H, 4H, and 15R polytypes are referred to as alpha-SiC.

Large numbers of experiments have been conducted by various researchers in an attempt to define stability regions for the basic silicon carbide polytypes. Work by both Knippenberg [5] and Inomata et al. [6] resulted in stability diagrams. In both studies, the stability was shown to increase with temperature moving from 3C to 2H to 4H to 15R to 6H. Specifically, the work by Inomata et al. indicates that 3C polytype will convert to a hexagonal polytype above about 1600°C and that near complete conversion, to the 6H polytype, will occur above approximately 2200°C. Thus, at high process temperatures, the formation of the 6H polytype is predicted.

A typical processing temperature for RBSC is 1500°C. Therefore, it is possible to produce RBSC ceramics with starting SiC powders of varying polytype, and maintain the polytype through the process. The present work produces RBSC ceramics with both 3C and 6H starting SiC powders. The ceramics are then characterized to assess the effect of polytype on the properties.

As stated, the finished RBSC ceramic in each case is a Si/SiC composite. For comparison and reference, the properties of monolithic Si and SiC are provided in Table I.

Table I: Properties of Silicon and Silicon Carbide for Comparison (primarily α-SiC)

Property	Si [Ref.]		SiC [Ref.]		Units
Density	2.33	[7]	3.21	[8]	g/cc
Young's Modulus	113	[7]	475	[8]	GPa
Poisson's Ratio	0.2	[9]	0.2	[8]	
Bulk Modulus	44	(calc)	198	(calc)	GPa
Shear Modulus	86	(calc)	264	(calc)	GPa
Thermal Conductivity	84	[7]	490 (300K) (α) 250-300* (β)	[10] [11,12]	W/m K
Coefficient of Thermal Expansion	2.6(300K)	[13]	2.3 (293K)	**	10^-6/K
Specific Heat	0.709	***	0.667(α) 0.669(β)	***	J/g K
Fracture Toughness	--		4.6	[8]	MPa m^(1/2)
Flexural Strength	62	[7]	228-261	[8]	MPa

*Temperature not provided in reference
**Measured by M Cubed Technologies, Inc.
***Calculated from condensed phase thermochemistry data for SiC and Si found at http://webbok.nist.gov

EXPERIMENTAL PROCEDURES

The samples generated for this paper were all produced in the following fashion. Firstly, a slip was developed with silicon carbide powder and a carbon based binder. The SiC powder type, 6H or 3C, and carbon content were systematically varied in the slip formulation. Secondly, a preform was cast from this slip. Finally, the preform was subjected to a reactive infiltration step in which the preform was brought into contact with a molten silicon bath under vacuum.

Three different sets of samples were made with varying types of SiC. Two of the silicon carbide powders used to make the samples were primarily the 6H polytype. The third silicon carbide powder used was primarily of the 3C polytype. Of the two powders with the 6H polytype, one has a nominal diameter of 50 μm, and the other has a nominal diameter of 10 μm. The primarily 3C polytype powder has a nominal diameter of 19 μm. Compositions of the three SiC powders with respect to the different polytypes are summarized in Table II.

Table II: Starting Powder Constituents

Material Description	XRD (Wt. %)					
	SiC (6H)	SiC (3C)	SiC (2H)	SiC (4H)	SiC (15R)	Graphite
6H 50 μm	98.8				1.2	
6H 10 μm	93.3			3.4	3.3	
3C 19 μm		96.5	0.3			3.2

The three sets of experimental samples made from these raw materials consisted of two monomodal SiC distributions and one bimodal SiC distribution. One of the two monomodal distributions contained the 6H 10 μm powder while the other contained the 3C 19 μm powder. The bimodal distribution contained both sizes of the primarily 6H powder. The 50 μm powder was the primary constituent. These will be referred to by the primary constituent and a number to indicate separate samples, which respectively becomes 6H 10 μm, 3C 19 μm, and 6H 50 μm.

Physical and mechanical properties were determined using the various methods described below. Density was measured by the water immersion technique, which can be found in ASTM Standard B 311. Poisson's ratio and Young's modulus were determined via the pulse echo technique described in ASTM Standard E 494-05. ASTM Standard C 1421 describes the process by which the fracture toughness of these samples was determined. The fracture toughness experiments were performed using a four-point chevron notch setup. Flexural strength tests were also run, with ASTM Standard C1161 used as a reference. For both the flexural strength and fracture toughness property measurements, a Sintech universal testing frame was utilized in conjunction with TestWorks materials testing software to perform both sets of experiments. Polytype compositions were determined with quantitative x-ray powder diffraction (XRD) techniques, and the microstructures of the samples were compared using a Leica D 2500 M optical microscope and the Clemex Vision PE imaging software.

The XRD measurements were performed on a Philips PW1800 diffractometer using Cu radiation at 40KV/30mA. Scans were done with a step size of 0.02° over a range of 10° to 70°. The counting time was ten hours. Upon collection of the diffraction patterns, the "Powder Diffraction File" of the International Centre for Diffraction Data was used to identify the crystalline phases. The "Rietveld Refinement Method" was used to determine the quantitative phase data[14,15].

Thermal properties were also measured and calculated. The thermal conductivity was calculated from Equation 1 below:

$$\lambda = \kappa \cdot C_p \cdot \rho \qquad (1)$$

where λ is the thermal conductivity, κ is the diffusivity, C_p is the specific heat, and ρ is the density. The diffusivity was determined using the laser flash technique per ASTM E 1461-01. The specific heat was found by differential scanning calorimetry using a Perkin-Elmer DSC 7.

RESULTS AND DISCUSSION
Microstructure and Composition

Microstructures of representative samples are shown in Figure 1. The lighter phase is Si and the darker phase is SiC. When compared to the images on the right, the images on the left represent samples with lower amounts of initial carbon. Greater amounts of carbon allow for more reaction during the molten Si infiltration step, and thus a higher final SiC content.

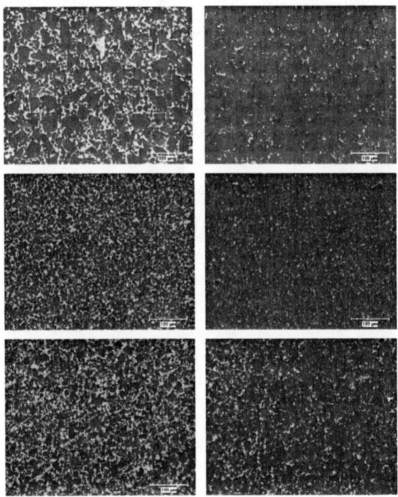

Figure 1: 50um 6H Sample (top), 10um 6H Sample (middle), and 19um 3C Sample (bottom)
(left image – low carbon ratio and right image – high carbon ratio)

X-ray diffraction (XRD) results are provided in Table III in weight percent. Table IV provides results of Si content in volume percent from three different measurements, namely (1) calculation from density, (2) measured by XRD and converted from weight to volume percent, and (3) measured by quantitative image analysis (QIA). In all cases, the amount of carbon in the starting preforms, with similar starting SiC weights, increases as the sample number increases.

Table III: Sample Composition based on XRD results

Labels	XRD (Wt. %)					
	SiC (6H)	SiC (3C) aka beta	SiC (2H)	SiC (4H)	SiC (15R)	Si
6H 50um 1	80.9	--	--	2.3	3.3	13.5
6H 50um 2	79.8	--	--	5.3	3.9	11.0
6H 50um 3	83.7	--	--	1.1	5.5	9.7
6H 50um 4	86.2	--	--	1.1	6.6	6.1
6H 50um 5	89.5	--	--	1.8	3.6	5.0
6H 10um 1	76.5	--	--	2.2	2.4	18.9
6H 10um 2	78.4	--	--	2.7	1.9	17.0
6H 10um 3	79.0	--	--	3.1	4.1	13.8
6H 10um 4	82.2	--	--	2.2	2.2	13.4
6H 10um 5	81.6	--	--	3.7	2.5	12.2
6H 10um 6	84.5	--	--	3.2	2.7	9.6
6H 10um 7	85.6	--	--	2.2	2.7	9.5
3C 19um 1	0.6	75.2	--	0.2	1.5	22.5
3C 19um 2	--	78.1	0.5	--	--	21.4
3C 19um 3	1.2	81.5	--	0.3	0.6	16.4
3C 19um 4	--	84.0	0.3	--	--	15.5
3C 19um 5	--	84.7	1.0	--	--	14.3
3C 19um 6	0.6	86.3	--	0.1	1.1	11.9
3C 19um 7	0.5	86.7	--	0.6	0.5	11.7

Table IV: Density, QIA and XRD Comparison of Silicon Phase

Label	Vol % Si from Density	Vol % Si from XRD	Vol % Si from QIA
6H 50um 1	22	18	20
6H 50um 4	12	8	7
6H 10um 1	30	24	26
6H 10um 3	23	18	20
6H 10um 5	18	16	14
6H 10um 6	15	13	13
3C 19um 1	36	29	32
3C 19um 3	28	21	26
3C 19um 6	22	16	24
3C 19um 7	19	15	15

The finished Si/SiC ceramics contain a mixture of the original SiC particles in the preform, reaction formed SiC (from Si + C → SiC reaction), and residual Si. The results in Table III demonstrate that the predominant SiC polytype in the composites is that of the starting SiC powder in the preform

That is, when the principley 6H or 3C powders were used in the preform, they remaine in the same polytype, and the polytypes of the reaction formed SiC matched the starting powde contents. Seeding is the most likely explanation for this, since other polytypes were n discovered in these samples. Furthermore, the 2H, 4H, and 15R polytypes were found in many the samples as well. While minor amounts of these polytypes may have been formed as the rest of the Si and carbon reaction during infiltration, it is much more likely that the bulk of the polytypes were initially present in the starting powder as evinced in Table II.

The thermal stability and transformation tendencies of these polytypes support tl hypothesis that the polytypes were either initially present or reaction formed, and n transformed. Bootsma et al. [16] provide experimental data for the 2H → 3C transformation 1500°C, 1600°C, and 1700°C. From this experimental data, it becomes apparent that a ten ho hold time at 1500°C is required in the transformation of 5% 2H to 3% 2H with the 3C polyty] making up the difference. The XRD data provided in Tables II and III give minor amounts, le than 7% by weight, for the 2H, 4H, and 15R polytypes. These should take significant amounts time, greater than 10 hours, at temperature to convert to the 6H polytype. For referenc transformation rate data from Bootsma et al. are provided in Table V.

Table V: Transformation Temperatures with Approximate Rate Constants [16]

Transformation	Temperature (°C)	Rate Constant (min⁻¹)
2H-->3C	1500	10^{-3}
3C-->6H	2000	10^{-3}
4H-->6H	2600	10^{-3}
15R-->6H	2300	10^{-3}

Note: Temperature and rate constant data estimated from chart

From these data, transformations of one polytype to another are not expected with respect to tl processing conditions to which the samples were exposed (nominally 1500°C for 2 hrs. und vacuum). This is proven out by experimental observations where significant changes of the 2] 4H, and 15R polytypes were not observed.

The XRD results in Table III show that Si/SiC ceramics with a wide range of Si conter can be produced by varying the silicon carbide to carbon ratio of the preform. Again, previously stated, within each set of data (6H 50μm, 6H 10μm, and 3C μm) the higher samp numbers refer to higher starting carbon content and ultimately greater amounts of SiC..

As shown in Table IV, the volume percent silicon measured by the XRD and Ql methods are relatively consistent. The volume percent calculated from the measured density w higher than either of the other two methods in all cases. As with the data in Table III, the trend decreasing silicon was seen with respect to higher carbon to silicon carbide ratios for ea method.

Physical and Mechanical Properties

A chart of density and Young's modulus as a function of volume percent SiC is provid in Figure 2. The density follows the typical additive rule of mixtures. Over this range of S volume percent, the Young's modulus also appears to follow a linear relationship. When plott in relation to the Hashin-Shtrikman bounds, however, a slight parabolic shape is observed, shown in Figure 3. These bounds were calculated based on the equations provided by Hashin al. [16] for Young's modulus.

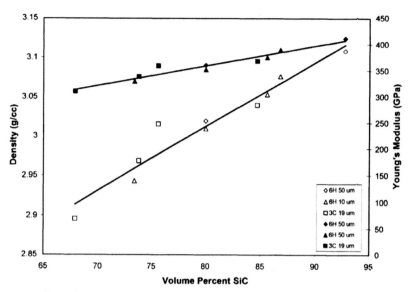

Figure 2: Variations of Density and Modulus as a function of Volume Percent SiC
(open symbols give density data, closed symbols give modulus data)

The primary polytype contained in these samples, whether it be 3C or 6H, brings no bearing upon the relationship between Young's modulus and density. As expected, the data in Figure 3 run along the upper bound. This is normally found when the material with higher elastic moduli is encapsulating the material of the smaller elastic moduli. If the data followed the lower bound, the material with the lower elastic moduli would be encapsulating that of the higher elastic moduli material [17, 18]. Due to the reactive nature of the RBSC process, the higher elastic moduli material, that being the SiC, form a continuous network. As the silicon reacts and wicks through the green preform, more silicon carbide is formed which creates this continuous network of silicon carbide as depicted in Figure 1.

Figure 4 provides the flexural strength and the fracture toughness as a function of volume percent SiC. The flexural strength is expected to increase with decreasing grain size assuming the Orowan relation where the strength is proportional to the grain (particle) diameter$^{(-1/2)}$ [19]. Although the differences are small, a trend of increasing strength with decreasing particle size is observed. Little effect of SiC:Si ratio on strength is shown in the data. From the current study, the fracture toughness is not affected by the predominant polytype present, the primary particle size used, or the volume percent SiC.

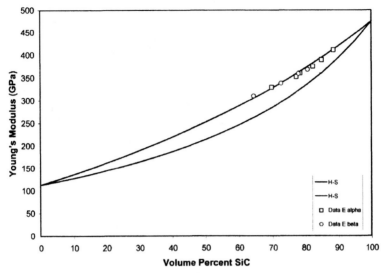

Figure 3: Variation of Young's Modulus as a Function of Volume Percent SiC.
The calculated H-S bounds are also plotted. (circles are 3C polytype and squares are 6H polytype)

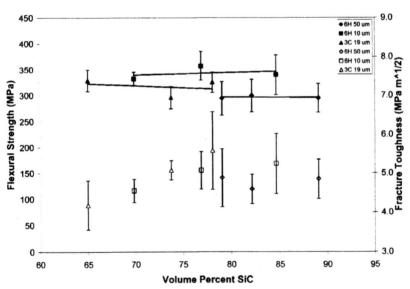

Figure 4: Flexural Strength and Fracture Toughness as a function of Volume Percent SiC
(closed points are flexural strength and open points are fracture toughness)

THERMAL PROPERTIES

An additional study involved the affect of the different primary polytypes, namely the 6H and 3C polytypes, on the thermal properties of these materials. The thermal properties include the specific heat, the thermal diffusivity, the thermal conductivity, and the coefficient of thermal expansion. Figure 5 provides data for the specific heat and thermal diffusivity of several samples.

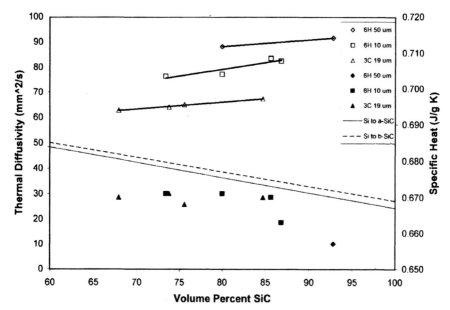

Figure 5: Thermal Diffusivity (open points) and Specific Heat Data (closed points)
(the solid line is theoretical for Si to α-SiC specific heat,
and the dashed line is theoretical for Si to β-SiC specific heat)

Comparing the literature data, the 3C polytype should have a slightly higher specific heat, Table I. From the data presented here, several cases stand out where the specific heat values are roughly the same for similar amounts of SiC. With the actual difference of the pure silicon carbide polytypes being 0.002 J/ g K, differences in the actual data are most likely contained within the experimental error of this measurement. The specific heat should therefore not have a great impact on the thermal conductivity of these materials with respect to their respective polytypes.

The thermal diffusivity values on the other hand do seem to be affected by the different polytypes. The particle size can be neglected as the 6H polytype samples made with both the 10 µm size and the primarily 50 µm size have greater diffusivity values than the 3C polytype samples made with a nominal diameter of 19 µm. This follows in the thermal conductivity values as well and is shown below in Figure 6. As shown in Table I, the literature value for thermal conductivity is much higher for α-SiC in comparison with the value provided for β-SiC. Fundamentally, the thermal conductivity of a material, whether it is α-SiC or β-SiC, is determined by the crystal structure and the electron and phonon mobilities.

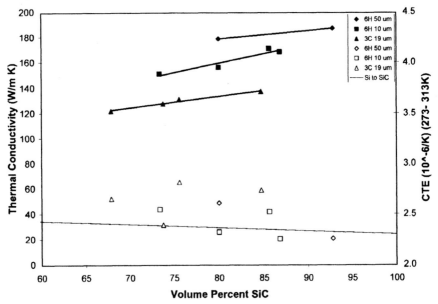

Figure 6: Thermal Conductivity (closed points) and CTE (open points) vs. Vol % SiC (the line is drawn between the CTE values for Si and SiC in Table III)

Clearly there is an upward trend in the data for thermal conductivity as the volume percent SiC increases. This is to be expected as the thermal conductivity of pure silicon carbide is much higher than that of pure silicon (Table I). The size effects also are seen. Apparent from Figure 6 is the primarily large size 6H polytype samples having a higher thermal conductivity than the smaller size 6H polytype samples. This may be attributed to the higher surface area involved in the samples fabricated from a smaller grit size; thus, the samples of a smaller grit size will have a greater area of interfaces where the thermal conductance is lower. A corollary argument can be made in this case with the help of the work of Molina et al. [20] in which they describe Al-SiC monomodal composites of increasing size from 6 to 130 μm with increasing thermal conductivity from 154 to 221 W/m K. While this is a different matrix phase, the response should be similar since both aluminum and silicon have lower thermal conductivity values than silicon carbide.

Technically, the thermal expansion of these materials should decrease with increasing SiC content based on the monolithic data of the thermal expansion data for pure silicon and SiC in Table 1. The aggregate data suggests a slight downward trend in thermal expansion with increasing SiC volume percent; however, this is not seen as statistically significant. Conclusions can not be drawn as to the effects of the silicon carbide polytype and its effect on thermal expansion. A broader range of volume percent silicon carbide may provide the sensitivity in which these differences can be assessed.

SUMMARY
A series of RBSC (Si/SiC) ceramics were produced and characterized. Variables were polytype of SiC, SiC:Si ratio and SiC particle size. There was a strong effect of SiC:Si ratio on Young's modulus, with higher SiC contents leading to higher Young's modulus values. Moreover, the data fell near the upper bound of the predictive Hashin-Stricktman bounds. There was little effect of SiC polytype on all properties examined, except for the thermal diffusivity which led to an effect on the thermal conductivity. With respect to thermal conductivity, ceramics made with beta (3C) polytype SiC consistently showed lower thermal conductivity than those made with alpha (6H) polytype SiC. By decreasing the particle size of SiC, a deleterious effect was noted on the thermal conductivity in that it decreased. The remaining properties measured did not demonstrate any significant particle size effects over the SiC content range under investigation.

REFERENCES
1. K.M. Taylor, "Cold Molded Dense Silicon Carbide Articles and Methods of Making the Same," U.S. Pat. No. 3 205 043, Sept. 7, 1965.
2. P.P. Popper, "Production of Dense Bodies of Silicon Carbide," U.S. Pat. No. 3 275 722, Sept. 27, 1966.
3. C.W. Forrest, "Manufacture of Dense Bodies of Silicon Carbide," U.S. Patent No. 3 495 939, Feb. 17, 1970.
4. J. Pezoldt, F.M. Morales and A.A. Kalnin, "Local Control of SiC Polytypes", *Phys. Stat. Sol.* **A204** [4] 1056-62 (2007).
5. W.F. Kinppenberg, *Phillips Research Reports,* **18** 161-274 (1963).
6. Y. Inomata, Z. Inoue and M. Mitomo, "Thermal Stability of 6H and 15R Types of SiC", *J. Crystal Growth,* **5,** 405-7 (1969).
7. *Metals Handbook: Desk Edition* (ASM International, Metals Park, OH, 1985).
8. *Engineered Materials Handbook, Vol. 4, Ceramics and Glasses* (ASM International, Metals Park, OH, 1991).
9. J. J. Wortman, R. A. Evans, "Young's Modulus, Shear Modulus, and Poisson's Ratio in Silicon and Germanium," *J. Appl. Phys.,* **36** [1] 153-156 (1965).
10. G.A. Slack, "Nonmetallic Crystals with High Thermal Conductivity," *J. Phys. Chem. Solids,* **34** 321-335 (1973)
11. http://www.cvdmaterials.com/literature/pdfs/CVD_SiC_Specs.pdf
12. http://www.morgantechnicalceramics.com/articles/CVD_SiC.htm
13. *ASM Handbook, Volume 2 Properties and Selection: Nonferrous Alloys and Special-Purpose Materials* (ASM International, USA, 1990).
14. H.M. Rietveld, "The Crystal Structure of some Alkaline Earth Metal Uranates of the Type M_3UO_6*," *Acta Cryst.* **20** 508-513 (1966).
15. H.M. Rietveld, "A Profile Refinement Method for Nuclear and Magnetic Structures," *J. Appl. Cryst.* **2** 65-71 (1969).
16. G.A. Bootsma, W.F. Knippenberg, G. Verspui, "Phase Transformations, Habit Changes and Crystal Growth in SiC," *J. Crystal Growth,* **8** 341-353 (1971).

17. Z. Hashin, S. Shtrikman, "A Variational Approach to the Theory of the Elastic Behaviour of Multiphase Materials," *J. Mech. Phys. Solids*, **11** 127-140 (1963).
18. M. K. Aghajanian, R. A. Langensiepen, M. A. Rocazella, J. T. Leighton, C. A. Anderson, "The Effect of Particulate Loading on the Mechanical Behaviour of Al_2O_3/Al Metal-Matrix Composites," *J. Mat. Sci.*, **28** 6683-6690 (1993).
19. W. D. Kingery, H.K. Brown, W.R Uhlmann, *Introduction to Ceramics Second Edition* (Wiley-Interscience, 1960).
20. J. M. Molina, et al., "Thermal Conductivity of Al-SiC Composites with monomodal and bimodal particle size distribution," *Mater. Sci. & Eng.*, **A** (2007) in-press.

IN-SITU REACTION SINTERING OF POROUS MULLITE-BONDED SILICON CARBIDE, ITS MECHANICAL BEHAVIOR AND HIGH TEMPERATURE APPLICATIONS

Neelkanth Bardhan[a,*] and Parag Bhargava[a]

[a] Indian Institute of Technology (IIT) Bombay, Powai, Mumbai – 400 076, India.
* Junior Year Undergraduate Student, Corresponding author.
Tel.: +91 – 09867170671; E-mail: z5d11018@iitb.ac.in
Mumbai, Maharashtra, India.

ABSTRACT

The purpose of this work was to study reaction bonding technique as an effective way to produce porous silicon carbide ceramics. These can be tailored to have near-zero dimensional change while sintering, high service temperature under load and excellent thermal shock resistance. Sintered porous mullite-bonded silicon carbide was produced having an open porosity of 20-25%. XRD and SEM were used to confirm formation of the mullite phase and to study the sample microstructure. Rectangular bars of the material made using the slip-casting method were tested for thermal shock resistance as a function of the critical quench temperature difference the samples could withstand. Cyclic quench tests carried out from 1000 °C to 5 °C showed a loss in elastic (Young's) modulus of the order of 40% of the initial value in the first 6-8 cycles. Further thermal cycling did not have a significant influence on the elastic modulus. The possibility of the use of porous mullite-bonded silicon carbide in various metal-melting/casting applications has been investigated.
Keywords: mullite, silicon carbide, porous materials, thermal shock resistance

INTRODUCTION

In recent years there has been tremendous interest in porous ceramics because of their applications as filters, membranes, catalytic substrates, thermal insulation, gas-burner media and refractory materials. These are due to their superior properties, such as low bulk density, high permeability, high temperature stability, erosion/corrosion resistance and excellent catalytic activity. One branch of this field is porous SiC ceramics, owing to their low thermal expansion coefficient, high thermal conductivity and excellent mechanical properties. However, it is difficult to sinter SiC ceramics at moderate temperatures due to their covalent nature.[1] In order to realize the low temperature fabrication of porous SiC ceramics, secondary phases may be added to bond SiC. Oxidation bonded porous SiC ceramics have been found to exhibit good thermal shock resistance owing to the microstructure with connected open pores.

A method of potentially avoiding the problems associated with the internal stresses caused when attempting to densify a ceramic containing rigid inclusions is by the addition of a material which will, during heat treatment, transform to produce a large volume expansion. This method of reaction sintering can be used to form a homogeneous ceramic material from a multi-component starting material. The application of this fabrication method is capable of reducing the overall volume change during sintering to as near to zero as possible.[2]

The present study aims at investigating the Reaction Bonded Silicon Carbide (RBSC) process to produce porous mullite-bonded SiC ceramics. Wu and Claussen (1991)[3] reported a technique to produce mullite ceramics starting from Al, SiC and Al_2O_3 powder mixtures. However for the purpose of this study it was decided to use only SiC and Al_2O_3 as the precursor powders; with SiC as the major component; so that after completion of the reaction the microstructure would be SiC bonded with mullite phase, with no residual alumina. This material was then tested for its mechanical properties like Young's modulus, Modulus of Rupture. Properties of Silicate-based SiC refractories have been reported to a certain extent by Reddy[4] and others. Its potential use as a refractory material has been evaluated by measuring its thermal shock resistance. A sample refractory that has been designed in the form of a crucible was used to try melting of non-ferrous metals in an induction furnace, followed by casting.

EXPERIMENTAL PROCEDURE

Raw Materials Used
(i)   Alumina powder, Al_2O_3, grade: MR01, supplier: Hindalco, Belgaum, India, $d_{50}=0.7\mu m$, BET surface area = $7m^2/gm$.
(ii)   Silicon Carbide powder, SiC, Grade: Bonded, Grit: F600, supplier: Carborundum Universal, Ernakulam, Kerala, India. BET surface area: $0.31m^2/gm$. Particle size (as estimated from BET surface area, and also observed in SEM micrograph) $\approx$ 6 μm.

Figure 1. SEM of as-received SiC powder at x1.0k, showing an average particle size of 5 to 10 μm.

(iii)   Ammonium polyacrylate dispersant, NH_4PAA, grade: M65, supplier: Aquapharm, Pune, India

Colloidal Processing and Consolidation
        The $SiC:Al_2O_3$ ratio was decided to be kept at 2.1:1 (vol. ratio of the total solids loading). It is generally accepted that the SiC powder surface consists of an oxide film and the major functional group on the surface of SiC powder is silanol group.[5] The surface charging of SiC powder in water is attributed to the dissociation of the silanol group according to the following reactions:

$$[SiOH_2]^+ \xleftarrow{\;H^+\;} [SiOH] \xrightarrow{\;OH^-\;} [SiO]^- + H_2O \qquad (1)$$

The addition of a strong organic base like Tetra Methyl Ammonium Hydroxide (TMAH) can promote the dissociation of the silanol group, inducing more negative charge to SiC surface. The stability of SiC suspension in an aqueous medium can be controlled by adjusting the pH of the medium. The optimum dispersion is observed at the pH value of 10, as revealed by the minimum in viscosity and by the maximum in zeta potential for this suspension as reported in the literature[6]. Since the alumina slurries themselves have a pH in the range of 9-10, for the purpose of this study it was decided not to use any additional dispersant for the SiC powder. The pH would be naturally adjusted away from the IsoElectric Point (IEP) of SiC, which is in the range of 2-3[6]. After 24 hrs. of milling pH of the slurry was measured to be 10.17 (pH meter, Eutech Instruments) thus suggesting that the slurry may be well-dispersed as per expectations.

To determine the optimal solids loading of the slurry, four different slurry compositions with 30, 40, 50, 60 vol.% solids loading were prepared. Alumina powder was added to distilled water along with ammonium polyacrylate dispersant. After milling it for 6hrs. using zirconia balls as grinding media (powder : grinding media = 1:1) SiC powder was added, and further milled for 18hrs., in polypropylene bottles on a pot-mill rotating at 70rpm. Stability of slurries was characterized by examining the settling behavior. For each of the as-milled slurries 25 ml. was poured into identical measuring cylinders. These slurries were allowed to stand undisturbed for a period of 24hrs. and the settling behavior of the 4 slurries was compared at regular intervals. A rheology study of the as-milled slurry was also done, on a CCMS type Rheometer (R/S plus, Brookfield, Maryland, USA). The slurry was subjected to a shear rate of 0-500 s^{-1} and the change in viscosity was measured.

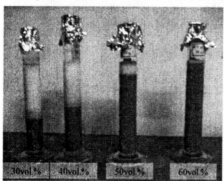
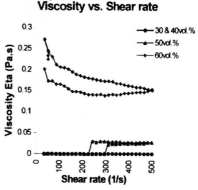

Viscosity vs. Shear rate

Figure 2. (a) Settling Behavior of slurries, after standing for 24 hrs. (b) Rheology of as-milled slurries.

It was decided to use 60 vol. % slurry for all future work as it was observed to be the most stable over a period of 24hrs., and showed shear-thinning behavior. The slurry was sufficiently fluid to be poured into the molds.

For the purpose of testing mechanical properties after sintering, it was decided to slip-cast bars using this slurry. The Plaster-of-Paris (PoP) powder used to make the molds was 'Khatri's Level Plaster of Paris', Mumbai, India. Based on results from prior work[7] it was decided to use a water : Plaster-of-Paris (PoP) powder ratio of 80ml/100gm.

The mold for casting bars was made by joining together microscope glass slides as the 4 vertical sides, a PoP slab as the base and an open top to pour the slurry (Figure 3); to achieve a uni-directional casting by draining water from the bottom of the mold. Bars having dimensions 75mm X 10mm X 10mm were cast.

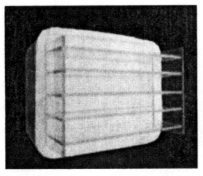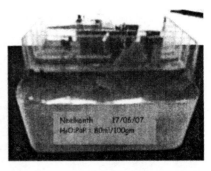

Figure 3. (a) Top view of glass-PoP mold (b) Slip-casting, using clay sealing to prevent leakage.

The as-milled slurry was poured into these molds. The thickness of the bars was controlled by the amount of slurry poured into the mold. To prevent water loss from the slurries by evaporation during the casting process, the top of the molds was tightly covered with aluminum foil. The green bars were then unmolded, dried in ambient conditions for 24hrs., followed by oven drying at 70-80°C for 12hrs. They were polished to required dimensions and sintered.

Sintering
The green bodies were sintered at 1550°C for 4hrs. Above 1400°C the mullitization reaction starts, and beyond 1500°C the reaction is almost complete and very little residual alumina is left.

Density measurement and Characterization
The density of the sintered bodies was measured using Archimedes method. Formation of the mullite phase was confirmed using X-Ray Diffraction (Philips PANalytical XRD machine, model no. Xpert Pro, using a Cu K- α radiation of 1.54Å) and Energy Dispersive Analysis of X-rays (Thermo NORAN X-Ray Fluorescence Spectroscopy Technique). A Hitachi S-3400N SEM was used to observe the microstructure of the material in the green, sintered states as well as to study the effect of thermal cycling on the microstructure.

Testing for Thermal Shock Resistance
In practical applications, porous SiC ceramics are usually applied under conditions of severe and repeated thermal shock. Thus, it is necessary to investigate the thermal shock behaviour of mullite-bonded porous SiC ceramics. The tests carried out to measure quantitatively the effect of repeated (cyclic) thermal shock.
The quench test was performed on two of the sintered bars, as representative samples, designated as 'Sample1' & 'Sample3', (10 ± 0.13 mm X 10 ± 0.13 mm in cross section). The notation Sample1 & Sample3 corresponds to two bars having a difference in geometry, primarily thickness; all other processing parameters being exactly the same for both. The thickness was varied in order to study the influence of sample size on the response to thermal shock treatment.
The peak temperature from which the samples were to be quenched was decided to be 1000°C (to ice-cold water at 5°C).

Procedure for thermal shock testing:

(i)   The as-sintered bars were placed in the pre-heated furnace (1000°C) and held at that temperature for 20min. to homogenize.

(ii) They were then quenched using tongs, in a bath of ice-cold water maintained at 5°C. The bath was chosen sufficiently large so that the heat withdrawal did not cause any appreciable rise in the temperature of the quenching medium.

(iii) The effect of the thermal shock on the mechanical properties bars was evaluated by measuring the Young's modulus of the bar. This was done using a non-destructive impulse excitation technique, which sets up vibrations in the sample that are detected and used to determine the Young's modulus. (Dynamic Elastic Properties Analyzer, Jagdish Electronics, Bangalore, India).

(iv) Steps 1 to 3 were repeated for several cycles and the change in the Young's modulus was compared with the initial (as-sintered) value.

Beyond a certain number of cycles there was stabilization in the mechanical properties. After the thermal cycling process was concluded the bars were measured for their Modulus of Rupture using a 3-point bend test method. with a span of 40mm between the supports and a crosshead speed of 0.5mm/min (Tinius Olsen (Surrey, England) machine, model H25KS). This value of MOR was compared to the value for as-sintered samples.

Refractory Applications: Metal Melting

To explore the possibility of using porous mullite-bonded SiC as a refractory material, it was decided to try metal-melting in crucibles made from this material. An induction furnace was used for this purpose. A cylindrical crucible having an outer diameter of 70mm and length 110mm was made by slip-casting in a split-type PoP mold, followed by sintering. Since SiC is a poor conductor of electricity, this was enclosed in a graphite crucible which acted as the susceptor for the induction furnace. The whole assembly was placed in the furnace, and metals tried out were Copper, Brass and Aluminum (Ferrous alloys were not explored because they require temperatures of the order of the sintering temp. of mullite-bonded SiC). The crucible was then inverted to cast the molten metal in molds. and the wettability of the mullite-bonded SiC with reference to these metals was observed.

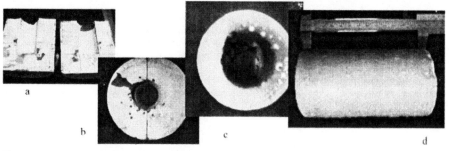

Figure 4. Mold for casting crucible (a) Split view (b) Top view; sintered crucible (a) Top view, showing glassy formation during the sintering process (b) Side view.

RESULTS AND DISCUSSION

Discussion of the Sintering Mechanism

According to the SiO_2–Al_2O_3 equilibrium phase diagram[8], mullite is formed from cristobalite and α- Al_2O_3 above 1400°C.

During the sintering process, the oxidation causes a weight gain for SiC. According to the literature, the oxidation of SiC particles occurs above 750°C[9] Before the beginning of the mullitization between oxidation-derived SiO_2 and α- Al_2O_3, the amorphous SiO_2 transforms partly to cristobalite. According to the XRD analysis of the SiC particles oxidized at different temperatures, the amorphous SiO_2 crystallizes to cristobalite above 1100°C.[9] Cristobalite can enhance the bonding between SiC particles and improve the strength of porous SiC ceramics. However, excessive cristobalite is harmful to the high-temperature properties of porous SiC ceramics because of its large coefficient of thermal expansion (17.5×10^{-6}/K at 20–700°C).[10]

A study of the SiC oxidation degree as a function of the sintering temperature during the fabrication of porous SiC ceramics by Ding et al.[11] shows that the oxidation degree increases from 11.2% to 33.4% with elevating the sintering temperature from 1400 to 1550°C. At the early stage of the SiC oxidation, O_2 diffuses to the surface of SiC easily and the oxidized products are SiO_2 and CO_2. As oxidation continues, the thickening of SiO_2 film and the formation of a mullite layer prevent or slow down the oxygen diffusion. This results in the lack of O_2 so that the oxidized products are gaseous SiO and CO. When these gases are given off, small pores are formed on the surface of SiC particles.

She et al.[12] have determined, using XRD, the major phases present in porous SiC ceramics sintered at different temperatures for 4 h. At 1400°C, porous SiC ceramics consist mainly of SiC, cristobalite and α- Al_2O_3, but slight mullite peaks can be found. When the temperature increases to 1450°C, the mullite peaks are obvious. At 1500 and 1550°C, the amount of α- Al_2O_3 decreases abruptly and more extensive mullitization occurs.

The mullitization between SiO_2 and α- Al_2O_3 can be explained by the solution precipitation mechanism.[13, 14] SiO_2 does not form a viscous liquid phase at 1400°C but shows viscous softening. Diffusion in the binary Al_2O_3-SiO_2 glasses can be viewed as a cooperative movement of oxygen-containing Al and Si complexes. With the movement of the Al-complex along the concentration gradient, there is a corresponding flux of an Si-complex toward the interface. These complementary movements can be visualized as viscous molecular masses of varying size moving by a cooperative rotation mechanism in which a minimum number of bonds must be broken. It is expected that the complexes which effect diffusion through the glass become progressively smaller with increase in concentration of Al_2O_3 and increase of temperature

Nucleation of mullite occurs at the Al_2O_3-SiO_2 interface. At 1450°C, the viscosity of SiO_2 glass decreases and hence diffusion through the viscous phase is enhanced, more so because the sizes of the diffusing complexes decrease. Above 1500°C, the mullite formation occurs by the reaction between cristobalite and α- Al_2O_3.

$$3Al_2O_3 + 2SiO_2 \Rightarrow 3Al_2O_3 \bullet 2SiO_2 \tag{2}$$

Because of the short diffusion distances achieved by viscous-flow-assisted sintering,[3] the rate of mullitization is accelerated drastically. After enough mullite is formed, the interfaces of SiO_2-mullite and Al_2O_3-mullite appear. SiO_2 and α- Al_2O_3 interdiffuse across the mullite layer and the formation of mullite is diffusion-controlled.

Discussion on the Dimensional Change during Sintering

Table 1: Dimensions of 4 representative bars, in their green and sintered states (all dimensions in mm)

Property	Sample 1		Sample 2		Sample 3		Sample 4	
	Green	Sintered	Green	Sintered	Green	Sintered	Green	Sintered
Length	74.44	74.96	74.62	75.00	71.00	71.46	75.00	75.48
Breadth	8.86	9.06	9.42	9.46	8.64	8.64	10.24	10.36
Thickness	7.92	8.00	8.38	8.52	6.18	6.18	9.30	9.40

Comparison of dimensions of the sintered and green bars indicates a change in dimension within +2% of the original, or in some cases, in the cross-sectional dimensions, there is no appreciable change.

A low shrinkage reaction-bonded mullite was investigated by Wu & Claussen (1991).[3] Additions of Al and SiC powders were made to Al_2O_3. Samples of varying compositions were isostatically pressed and sintered, with a 1200°C hold for oxidation of the SiC and a 1550°C hold for the mullite reaction. The addition of silicon carbide allows a reduction in sintering shrinkage through the large volume expansion of 108% involved during oxidation, and the smaller volume expansion of approximately 14% upon formation of mullite. By comparison, the oxidation of Al involves a 28% volume expansion. Wu & Claussen derived an equation which gives the approximate linear shrinkage, S, of a compact, related to the composition and state of the starting and finishing material:

$$S = \left(\frac{k\rho_0}{\rho}\right)^{\frac{1}{3}} - 1 \tag{3}$$

where ρ_0 and ρ are the relative green and sintered densities, and the constant k is given by the following equation:

$$k = \frac{0.2(1-f)V_{Al} + 1.32V_{SiC}}{1 + 0.28fV_{Al}} + 1 \tag{4}$$

where V_{Al} and V_{SiC} are the respective volume fractions of aluminium and silicon carbide used, and f is the fraction of Al oxidised during attrition milling.

If SiC alone is used, as in the present study, the previous equation reduces to:

$$k = 1.32V_{SiC} + 1 \tag{5}$$

$$\therefore S = \left[\frac{(1.32V_{SiC} + 1)\rho_0}{\rho}\right]^{\frac{1}{3}} - 1 \tag{6}$$

If S is set to zero, i.e. the requirement of zero overall shrinkage is applied, the volume fraction of SiC necessary is then set by the following:

$$V_{SiC,0} = \frac{1}{1.32}\left(\frac{\rho}{\rho_0} - 1\right) \tag{7}$$

Therefore, to minimise the volume fraction of silicon carbide required for zero shrinkage, the green density ρ_0 is to be made as high as possible. Using a mixture of SiC, Al and Al_2O_3, isostatically pressed at 380 MPa to 68.5% of theoretical density, Wu & Claussen obtained a linear shrinkage after sintering of 0.1%.

In the present study the near zero-shrinkage phenomenon was observed in yet another instance. The crucible made for metal melting, was measured to have an outer diameter of 70mm in the green state, and 68.90mm in the sintered state.

Characterization
The density of the sintered bars was measured using Archimedes method, and was determined to be 2.47 gm/cc with a Standard deviation = 0.01614 gm/cc, Coeff. Of variance = 0.6531%
Open porosity was determined to be (on average) 24%.
This value of density obtained is actually better than a comparable commercial Refractory tube, sheath or rod made of mullite-bonded porous SiC, Anderman Ceramics[1], having a composition
SiC = 90, Al_2O_3 = 6.5 & SiO_2 = 3.5 wt.%, density = 2.2gm/cc, open porosity = 25%
X-ray Diffraction was used to confirm the formation of the mullite phase:

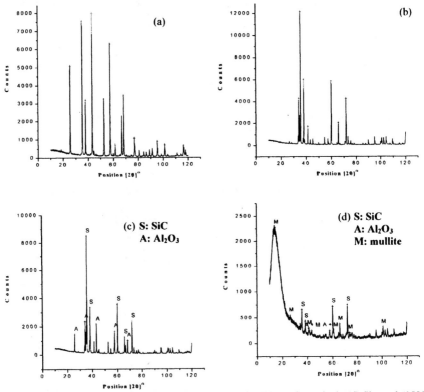

Figure 5. XRD patterns for the raw materials (a) Al_2O_3 (b) SiC (c) Green body (d) Sintered (1550°C for 4hrs.) body. The characteristic mullite peaks are clearly observed at 2θ = 26, 35, 61 & 66°

Microstructural Characterization

Figure 6. SEM of slip-cast green body at (a) x500; (b) x1.5k, showing uniform coating of the finer Al_2O_3 particles on the coarser SiC particles.

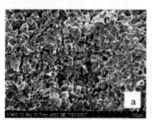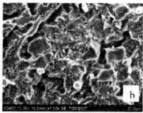

Figure 7. Fracture surface of as-sintered sample, at (a) x500; (b) x1.5k; (c) x5k; showing a porous microstructure.

Energy Dispersive Analysis of X-Rays (EDAX) was performed to ascertain the composition of the layer-like formation on the SiC grains (point 2) and the needles (point 1)

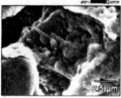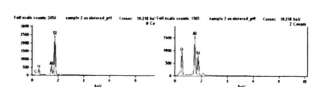

Figure 8. EDAX on as-sintered sample.

	Atom %		
	C	O	Al	Si
Sample 2 as sintered_pt1	30.05	29.43	6.67	33.85
Sample 2 as sintered_pt2		57.06	25.09	17.85

Pt. 2 suggests formation of a mullite phase surrounding the grain, since mullite ($3Al_2O_3.2SiO_2$) has an atom ratio Si:Al:O = 9.5%:28.6%:61.9%. The formation of the mullite phase was confirmed by XRD. The needles indicated by pt. 1 are most probably SiO_2.

To determine the uniformity in the porous microstructure, as-sintered sample was polished using 1μm diamond paste, and observed under optical microscope.

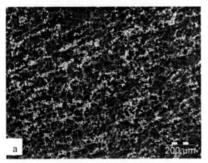

Figure 9. Optical micrographs of polished surfaces of as-sintered sample, at (a) x100; (b) x500 showing a uniformly porous microstructure.

The diamond-polished sample, when observed in SEM, gave a clear image of the distribution of the SiC grains in the as-sintered state. EDAX performed at the red location indicated in Fig. 10 (b) confirmed the grain to be SiC. This sample was then used to determine the compositional variation across the grain, by performing an X-Ray linescan.

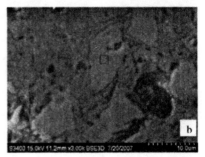

Figure 10. SEM images of polished surface of as-sintered sample at x3.0k, using (a) secondary electrons (b) back-scattered electrons.

The linescan (Fig. 11 (a)) proves that inside the grain, the composition of SiC is higher, while outside, the composition is dominated by Al_2O_3, with a relatively lower percentage of Si. This is confirmed by the dot map (Fig. 11(b)) shown alongside. This is again indicative of the formation of a mullite phase surrounding the SiC grains.

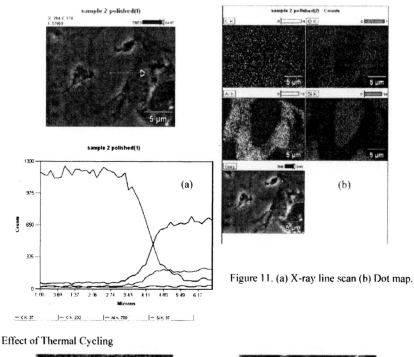

(a)

(b)

Figure 11. (a) X-ray line scan (b) Dot map.

Effect of Thermal Cycling

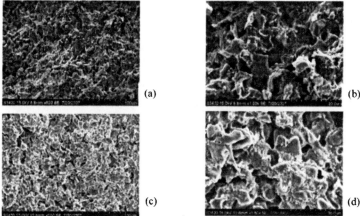

Figure 12. SEM images of fracture surface after thermal cycling, for (1) 6 cycles, at (a) x500, (b) x1.5k; (2) 10 cycles, at (c) x500, (d) x1.5k. Comparison with the as-sintered microstructure at the same magnifications (Fig. 7 a, b) shows considerable grain coarsening with repeated thermal cycling.

To demonstrate the change in composition after thermal cycling, an EDAX was performed inside one of the grains of SiC:

sample3_10cycles

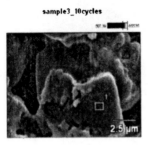
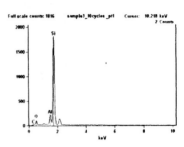

Figure 13. EDAX on sample after thermal cycling.

| | Atom % | | | |
	C	O	Al	Si
sample3_10cycles pt1	15.58	17.97	5.59	60.85

It is evident that repeated thermal cycling has caused a considerable oxidation of the SiC to SiO_2, which is proved by the Si:C:O atom % ratio. The % of C is appreciably lower, and a significant peak has been detected for O, compared to the values for the as-sintered sample.

The change in the mechanical properties of the bar, as a consequence of thermal cycling, was evaluated by measuring the Young's modulus.

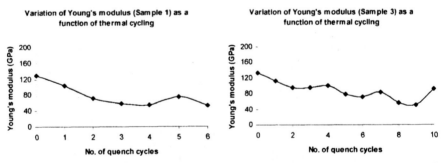

Figure 14. Young's modulus, measured by Impulse Excitation technique (Nondestructive) after repeated thermal cycling (quenching from 1000 °C to 5 °C): (a) 6 cycles (b) 10 cycles.

In general, Young's modulus decreases rapidly with cyclic thermal shock treatment. However, after about 6-8 cycles, the modulus stabilizes at about 40% of the initial value (as-sintered).

Measurement of Modulus of Rupture:

As-sintered (average of 4 samples): 63MPa; after thermal cycling: 10MPa. It is clear that thermal cycling causes a sharp fall in the flexural strength. This resembles with the data reported by Ding et al.[15]

CONCLUSION

The main objective of the project, ie. to study the reaction bonding technique as an effective way to produce porous silicon carbide ceramics, has been successfully achieved. This includes the colloidal processing of the precursor powders, compaction through slip-casting, and finally sintering to achieve the mullitization reaction. Characterization of the sintered and polished samples using SEM and XRD have confirmed the formation of the mullite phase.

The mechanical as well as the high temperature properties of this material, have been studied in some detail. Bars fabricated using the slip-casting technique have shown good density, Young's modulus values and flexural strength, which corroborate the values reported in literature. Cyclic thermal shock quench tests have been performed on the bars, to study the effect on mechanical properties. These observations have been supported by microstructural analysis.

The secondary objective of the project, ie. to study the various high-temperature applications of mullite-bonded SiC, needs to be further worked upon. Preliminary literature survey suggests a promising field of work for developing refractories using this material, having a use temperature of as high as 1450°C in air, and good resistance to corrosion by molten metals and gases. The fabrication of a crucible for metal-melting applications has been a step in this direction. Special efforts have to be put in this direction, for the purpose of designing a low-cost solution to study the high-temperature MOR of bars made of this material. This is a crucial step for successfully designing and producing refractories on a commercial scale.

ACKNOWLEDGEMENT

I would like to thank Indian Institute of Technology (IIT) Bombay, for funding this work under the Undergraduate Research Opportunity Program (UROP). I am grateful to the Physics Dept. of IIT Bombay for the use of the X-Ray Diffraction facility. I would like to acknowledge the creative inputs and suggestions of my seniors, Mr. Sabyasachi Roy and Mr. Ajaykumar Jena. Thanks are also due to Mr. Sujit Zagde and Mr. S. S. Dhoke of Powder Metallurgy Lab, IITB, for their help as lab assistants.

REFERENCES

[1]Riedel R., Passing G., Schonfelder H., Brook R. J., Synthesis of dense silicon-based ceramics at low temperatures, *Nature*, 1992, **355**(6362), 714-717.

[2]Pearce D. H., Fabrication and Evaluation of an Oxide-Oxide Ceramic Matrix Composite, Ph.D. thesis, Univ. of Birmingham, July 2007.

[3]Wu S. & Claussen N., Fabrication and Properties of Low-Shrinkage Reaction-Bonded Mullite, *J. Am. Ceram. Soc.* **74** (10) pp 2460-2463, 1991.

[4]Reddy N. K., Properties of Silicate-Bonded Silicon Carbide Refractories, *Materials Letters* **47** (2001) 305-307.

[5]Huang, Q., Gu, M., Sun, K. and Jin, Y., Effect of pretreatment on the properties of silicon carbide aqueous suspension, *Ceram. Int.* 2002, **28**(7), 747–754.

[6]Wei Li, Ping Chen, Mingyuan Gu., Yanping Jin, Effect of TMAH on rheological behavior of SiC aqueous suspension, *J. Eur. Cer. Soc.* **24** (2004) 3679–3684.

[7]Bardhan N., Roy S., Jena A. K. and Bhargava P., Influence of plaster of paris mold microstructure on sintered slip cast compacts, presented at International Seminar on Ceramics – CeraTec 2007, Vishakhapatnam, India.

[8]Kingery, Bowen & Uhlmann – Introduction to Ceramics, 2nd Edition, John Wiley, 1976.

[9]Ding, S., Zhu, S., Zeng, Y. and Jiang, D., Effect of Y_2O_3 addition on the properties of reaction-bonded porous SiC ceramics, *Ceram. Int.*, 2006, **32**(4), 461–466.

[10]Touloukian,Y. S., Thermophysical Properties of Matter, *vol. 13*. IFI/Plenum Data Company Press, New York, 1979.

[11]Ding S., Zhuc S., Zenga Y., Jiang D., Fabrication of mullite-bonded porous silicon carbide ceramics by *in situ* reaction bonding, *J. Eur. Ceram. Soc.*, 2007, **27**, 2095–2102.

[12]J.H. She, H. Schneider, T. Inoue, M. Suzuki, S. Sodeoka, K. Ueno, 'Fabrication of low-shrinkage reaction-bonded alumina–mullite composites', *Materials Chemistry and Physics* **68** (2001) pp 105–109.

[13]Davis, R. F. and Pask, J. A., 'Diffusion and reaction studies in the system Al_2O_3–SiO_2' *J. Am. Ceram. Soc.*, 1972, **55**(10), 525–531.

[14]Mechnich, P., Schneider, H., Schmucker, M. and Saruhan, B., 'Accelerated reaction bonding of mullite' *J. Am. Ceram. Soc.*, 1998, **81**(7), 1931–1937.

[15]Ding S., Zeng Y., and Jiang D., 'Thermal shock behaviour of mullite-bonded porous silicon carbide ceramics with yttria addition' *J. Phys. D: Appl. Phys.* **40** (2007) 2138–2142.

BIBLIOGRAPHY
[11]Data obtained from website
http://www.matweb.com/search/DataSheet.aspx?bassnum=CANDER06&ckck=1

STUDY ON ELASTO-PLASTIC BEHAVIOR OF DIFFERENT CARBON TYPES IN CARBON/CARBON COMPOSITES

Soydan Ozcan[a,b], Jale Tezcan[c], Jane Y. Howe[d], Peter Filip[a,b]
a. Mechanical Engineering and Energy Processes, Southern Illinois University, Carbondale, IL
b. Center for Advanced Friction Studies, Southern Illinois University, Carbondale, IL
c. Civil and Environmental Engineering, Southern Illinois University, Carbondale, IL
d. Materials Science and Technology Division, Oak Ridge National Laboratory, Oak Ridge, TN

ABSTRACT

The mechanical properties of carbon/carbon composites are sensitive to their crystallographic orientations and microstructures. Inelastic behavior can be characterized by studying their stress-strain behavior. Therefore, the aim of this work was to use nanoindentation methods to investigate the elasto-plastic behavior and related microstructures of PAN-fiber reinforced carbon matrix composites heat-treated at 2100°C. The microstructures were characterized using polarized light microscopy and high-resolution transmission electron microscopy. Nanoindentation tests were carried out to obtain loading-unloading cycles and elastic modulus data at different indentation depths using a three-faceted Berkovich-type diamond indenter. The residual displacement at complete unloading was correlated with the microstructure data to reveal the extent of the deformation of crystallites and graphene sheets.

The pitch fiber and rough laminar pyrocarbon exhibited plastic behavior that can be attributed to the low shear resistance due to weak bonding between the well-organized graphene sheets. On the other hand, the PAN fiber, charred resin and isotropic pyrocarbon all exhibited almost full elasticity within the applied range of indentation depths.

INTRODUCTION

Carbon fiber reinforced carbon matrix (C/C) composites have found wide application in commercial and civilian aircraft, recreational, industrial, and transportation markets. C/C composites are ideally suited to applications where high strength, high stiffness, low density, and fatigue resistance are critical requirements. In addition, they can be used when high temperature resistance, chemical inertness and high dampening are desired[1-3].

Microstructure-mechanical property relationships of carbon fiber and matrix have long been of interest to scientists seeking to improve the composite performance. The two major parts of C/C composites are the carbon fiber and the carbonaceous matrix, both of which usually have different microstructure and properties. The most commonly used carbon fibers are PAN, pitch and rayon based carbon fibers. Microstructure and mechanical properties of C/C composites show significant variation depending on the fiber and matrix type, and the processing method. Similarly, the microstructure of matrix atoms exhibits high variation with the type and processing conditions, which are considered the most important factors in determining the mechanical and thermal properties as well as frictional performance[4-7].

In this study, nanoindentation tests were used to examine the load-displacement behavior of different constituents of C/C composites, to obtain information on their hardness, elasticity, plasticity, and recovery

141

EXPERIMENTAL PROCEDURE

Two types of commercial C/C composites were used in this study: a two directional randomly chopped pitch fiber reinforced composite with charred resin/CVI matrix (CC-A21) and a three directional composite with needled felt ex-PAN fibers and CVI matrix (CC-D21). The C/C composites were kindly provided by Honeywell Aircraft Landing Systems. The as-received densified C/C composite samples were heat-treated at 2100°C in a graphitization furnace (Model: TP-4X10-G-G-D64A-A-27, Centorr Associates Inc) with protective argon atmosphere.

After the heat treatment, representative specimens with dimensions of approximately 15 mm x 15 mm x 15 mm were cut from different discs, mounted into an epoxy resin, ground and then polished using diamond polishing slurries with a grain size ranging from 6 to 0.25 μm. The polishing was completed with a 0.05 μm grain size alumina suspension.

The nanoindentation experiments were conducted at room temperature with a Nano Indenter® XP system (MTS Nanoinstruments, Knoxville, TN) using a Berkovich-type diamond tip. Before each test, the system was calibrated using a fused silica. The continuous stiffness mode (CSM) was used in the tests. Thirty randomly selected different fiber and CVI matrix locations were indented for each component of C/C composites. The method of Oliver and Pharr was employed for the elastic modulus calculations[8].

Thin foils for the HRTEM study were prepared from 300 μm thick sections using Linear Precision Diamond Saw, Isomet 4000. The 3 mm - diameter disks were core-drilled from these sections (Core Drill, VCR Group, Model V 7110) and dimpled down to a thickness of about 5μ (Dimpler, D 500i, VCR Group). The ion mill (Gatan, 691 Precision Ion Polishing System) was used for the final thinning and polishing stage with a beam at 4° angle at 3 keV. The HRTEM studies were carried out using a 200kV JEOL 2010 EF transmission electron microscope.

RESULTS AND DISCUSSIONS

Representative micrographs of the two types of C/C composites examined by PLM are given in Figure 1. Composite CC-A consists of chopped pitch-based carbon fiber embedded in charred resin matrix which was finally densified using CVI process (Figure 1a, b). CVI matrix in composite CC-A (Figure 1b) displayed a low extinction angle, indicating isotropic microstructure. Pitch fibers exhibited a strong reflectance suggesting a highly graphitizable microstructure (Figure 1b). The charred resin matrix showed mostly homogenous contrast to the polarizer activity. Around the pitch fiber and CVI matrix boundary, a different microstructure with slightly lighter colors were observed (Figure 1 a, b), indicating stress induced graphitization. The internal stress build-up caused by the large volume shrinkage during carbonization results in the local graphitization of the resin-based carbon matrix at high temperatures[9,10].

Typical PLM micrographs of CC-D composite consisting of PAN-based carbon fiber and CVI matrix are shown in Figure 1c and d. CVI matrix displayed a strong optical activity (Ae=20) signifying a rough laminar microstructure (Figure 1d). PAN fiber, having a turbostratic microstructure, did not exhibit any distinct response to polarizer activity.

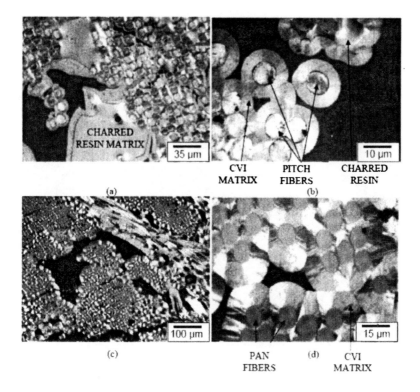

Figure 1. Typical low and high magnification polarized light micrographs of polished sections of CC-A21 (a, b) and CC-D21 (c, d).

Elastic moduli obtained for different structural constitutes of two composite types and reported values in literature are given in Table I. Elastic moduli were calculated from the load versus displacement data obtained by the nanoindenter[8]. Average values of the elastic moduli and standard deviations were obtained from 30 tests performed for each component in parallel and normal direction with respect to the C-fiber axis.

Table I. Elastic moduli obtained for different structural constitutes of two composite types and reported values in literature.

C/C component	Measured Elastic Modulus (GPa)	Elastic Modulus reported in literature
CC-A-Fiber (pitch) parallel	14.2 ± 1.1	30.31±1.19 (ref. 11)
CC-A-Fiber (pitch) normal	23.5 ± 1.2	15.30±0.31 (ref. 11) 14.18±0.32 (ref. 11) 14.23±0.28 (ref. 11) 16.11±0.37 (ref. 11)
CC-A-Matrix (CVI) (isotropic)	18.0 ± 2.1	30.69±0.97 (ref. 11)
CC-A-Matrix (Charred Resin)	24.3 ± 1.9	17.93±0.38 (ref. 11) 18.44±0.41 (ref. 11) 15. 24±0.37 (ref. 11) 21.04±0.55 (ref. 11) 16.30±0.34 (ref. 11)
CC-D-Fiber (PAN) parallel	29.5 ± 3.8	24.97±0.92 (ref. 11)
CC-D-Fiber (PAN) normal	17.2 ± 2.3	15.80±0.37 (ref. 11)
CC-D-Matrix (CVI) (rough laminar)	13.2 ± 4.1	9-12 (ref. 12)
Smooth laminar pyrocarbon (CVI)	N/A	39 (ref. 12)

Axial compression of the C-fiber causes two main deformation modes: kinking and shearing of the graphene sheets in the crystallites. Schematics of these modes are given in Figure 2. Kinking of the graphene sheets occurs in the crystallite and crystallite boundaries where lamellae recoil and form steps under a load. Even though kinking involves elastoplastic deformation, buckling of crystallites due to kinking of the graphene sheets, is mostly elastic. On the other hand, deformation due to shearing of crystallites or graphene sheets demonstrates plasticity. The deformation of pitch fiber involves significant plasticity, indicating strong prevalence of the shearing mode in the measured displacement.

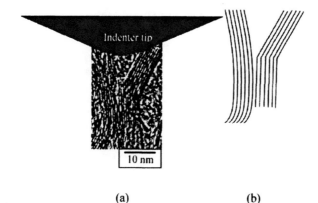

(a)   (b)

Figure 2. HRTEM micrograph of longitudinal section of PAN fiber (a). Schematic of basic crystallites microstructure seen in given HRTEM micrograph of longitudinal section of PAN fiber (b).

The average elastic moduli measured by the nanoindentation method, where the fiber is under axial compression, is 14.2 and 29.5 GPa for the pitch fiber and the PAN fiber, respectively. These values are much smaller than the values reported under tension, which typically are in the range 200-800 GPa [1, 2, 13-15]. The difference in the stress-strain behavior, as well as the failure mechanism, of the C-fibers in compression and tension can be explained by the anisotropic nature of the fiber. Tensile deformation in the PAN fiber starts with the initial straightening of ribbon shape crystallites. Tensile stresses start to accumulate when crystallites' movement is limited by the disclination in the microstructure. When the tensile stress reaches the critical value, rupture typically starts from the defects in the misoriented crystallites. The stress is transferred onto adjacent crystallites as the crack propagates [1,4,14]. As explained previously, compressive deformation of fiber involves different mechanisms. The values reported in this study are the compressive elastic moduli measured by nanoindenter and are comparable with the published data [11,12,15].

For matrices in both composites, the elastic modulus decreased with a better ordered carbon microstructure. Since the elastic modulus of rough laminar matrix (better ordered carbon) is 13.2 GPa, it is found to be 24.3 and 18.0 GPa for charred resin and isotropic pyrocarbon matrix (low-ordered carbon microstructure), respectively. Because better ordered carbon associated with the deformation due to shearing of crystallites or graphene sheets demonstrates plasticity.

Figure 3 through 6 present the results of the nanoindentation tests, where the samples were subjected to a single loading-unloading cycle. The inserts show the residual displacements upon complete unloading. The tests were controlled in such a way that unloading started after the preset maximum applied force 4.1 mN. The elastic moduli were calculated from the initial slope of the unloading curve.

Isotropic CVI matrix of CC-A21 exhibited the largest displacement (over 400nm) at the maximum load (Figure 3). The charred resin matrix with less organized turbostratic microstructure had a maximum deformation of only 275 nm (Fig. 4). The plasticity behavior was characterized according to the residual displacement at completed unloading, rather than the match of the loading and unloading branches. The pitch fiber exhibited the largest residual displacement, due to the low shear resistance between well-organized and preferentially oriented graphene sheets. This can be attributed to the stronger in-plane covalent bonds of the graphene sheets. On the other hand, the PAN fiber exhibited almost complete strain recovery (Figure 6), because of the turbostratic nature of the ribbon-shape crystallites. The deformation in the charred resin is mostly elastic, which can be explained by the high shear resistance of the turbostratic graphene sheets.

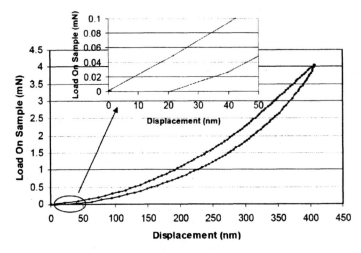

Figure 3. Load vs. displacement graph detected in nanoindentation experiment with CVI carbon matrix in CCA-21 Sample loaded in fiber axis direction.

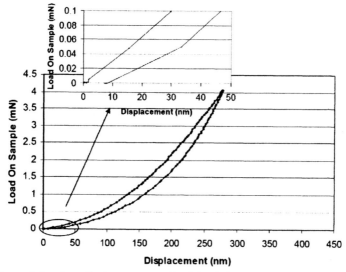

Figure 4. Load vs. displacement graph of charred resin matrix in composite CC-A21. Sample loaded in fiber axis direction.

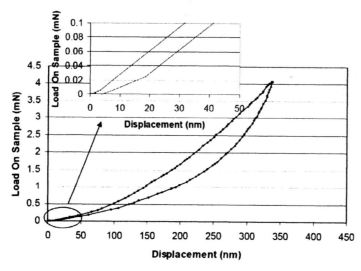

Figure 5. Load vs. displacement graph of pitch fiber in composite CC-A21. Sample loaded in fiber axis direction.

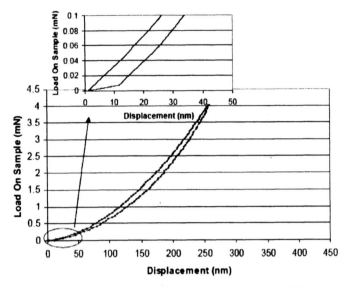

Figure 6. Load vs. displacement graph of PAN fiber in composite CC-D21. Sample loaded in the fiber axis direction.

CONCLUSIONS

Two main deformation modes were observed under axial compressive loading of the carbon fiber: shearing and kinking of graphene sheets in the crystallites. Deformation due to kinking is mostly elastic. On the other hand, shearing of crystallites or graphene sheets is associated with debonding of π (van der Waals) bonds which fail at low debonding energies, and thus, causes plasticity. In the pitch fiber and rough laminar pyrocarbon, the weak bonding between the well-organized graphene sheets results in low shear resistance, explaining the observed plasticity. On the other hand, the less-organized carbon, as in the PAN fiber, charred resin and isotropic pyrocarbon, exhibited almost full elasticity within applied displacement limits. This behavior can be attributed to the prevalence of the kinking mode in compressive deformation behavior.

ACKNOWLEDGEMENT

This research was sponsored by the National Science Foundation (Grant EEC 3369523372), State of Illinois and a consortium of 11 industrial partners of Center for Advanced Friction Studies (http://frictioncenter.siu.edu). The high-resolution TEM characterization was carried out at the Center for Microanalysis of Materials, University of Illinois, which is partially supported by the U.S. Department of Energy under grant DEFG02-91-ER45439. The authors acknowledge the contribution of Mr. Bijay Gurung for his help on nanoindentation studies, and Micro-imaging and Analysis Center at Southern Illinois University for assisting with the microscopy studies.

REFERENCES

[1]E. Fitzer, "PAN-Based Carbon Fibers--Present State and Trend of the Technology From the Viewpoint of Possibilities and Limits to Influence and to Control the Fiber Properties by the Process Parameters," *Carbon*, **27[5]** 621-645.1988

[2]M. Inagaki, "New Carbons: Control of Structure and Functions,": Elsevier. (2000).

[3]D. D. Edie, "Carbon Fibers, Filaments, and Composites, Edited by JL Figueiredo, et. al." In. Kluwer Academic Publishers, 1990.

[4]P. Delhaès, "Fibers and Composites,": Taylor & Francis. (2003).

[5]. Ozcan and P. Filip, "Microstructure and wear mechanisms in C/C composites," *Wear*, **259[1]** 642-650.2005.

[6]. Reznik and D. Gerthsen, "Microscopic study of failure mechanisms in infiltrated carbon fiber felts," *Carbon*, **41[1]** 57-69.2003.

[7]O. Siron, G. Chollon, H. Tsuda, H. Yamauchi, K. Maeda, andK. Kosaka, "Microstructural and mechanical properties of filler-added coal-tar pitch-based C/C composites: the damage and fracture process in correlation with AE waveform parameters," *Carbon*, **38[9]** 1369-1389.2000.

[8]W. C. Oliver and G. M. Pharr, "Improved technique for determining hardness and elastic modulus using load and displacement sensing indentation experiments," *Journal of Materials Research*, **7[6]** 1564-1583.1992.

[9]Y. Hishiyama, M. Inagaki, S. Kimura and S. Yamada , "Graphitization of carbon fiber/glassy carbon composites", *Carbon*, **12** 249–258. 1974.

[10]R.J. Zaldivar and G.S. Rellick , Some observations on stress graphitization in carbon–carbon composites. *Carbon* **29 [8]** 1155–1163. 1991.

[11]D. T. Marx and L. Riester, "Mechanical properties of carbon-carbon composite components determined using nanoindentation," *Carbon(UK)*, **37[11]** 1679-1684.1999.

[12]Diss, J. Lamon, L. Carpentier, J. L. Loubet, andP. Kapsa, "Sharp indentation behavior of carbon/carbon composites and varieties of carbon," *Carbon*, **40[14]** 2567-2579.2002.

[13]M. Savage, "Carbon-Carbon Composites,": Kluwer Academic Publishers. (1993).

[14]R. Thomas, "Essentials of Carbon-Carbon Composites," *Royal Society of Chemistry, Turpin Distribution Services Limited, Blackhorse Rd, Letchworth, Herts SG 6 1 HN, UK, 1993*. 234.1993.

[15]M. Kinari, K. Tanaka, S. Baba, andM. Eto, "Nanoindentation behavior of a two-dimensional carbon-carbon composite for nuclear applications," *Carbon*, **35[10]** 1429-1437.1997.

EFFECTS OF TEMPERATURE AND STEAM ENVIRONMENT ON CREEP BEHAVIOR OF AN OXIDE-OXIDE CERAMIC COMPOSITE

J. C. Braun, M. B. Ruggles-Wrenn*
Department of Aeronautics and Astronautics
Air Force Institute of Technology
Wright-Patterson Air Force Base, Ohio 45433-7765

ABSTRACT

The tensile creep behavior of an oxide-oxide continuous fiber ceramic composite (CFCC) was investigated at 1000 and 1100 °C in laboratory air and in steam. The composite consists of a porous alumina matrix reinforced with laminated, woven mullite/alumina (Nextel™720) fibers, has no interface between the fiber and matrix, and relies on the porous matrix for flaw tolerance. The tensile stress-strain behavior was investigated and the tensile properties measured. Tensile creep behavior was examined for creep stresses in the 80-150 MPa range. Primary and secondary creep regimes were observed in creep tests at all temperatures investigated. Creep rates increased with increasing temperature and creep stress. The presence of steam accelerated creep rates and dramatically reduced creep life. The detrimental effects of steam become progressively more pronounced with increasing temperature. At 1000 °C, creep run-out defined as 100 h at creep stress was achieved in all tests. At 1100 °C, creep run-out was achieved in all tests in air and only in the 100 MPa test in steam. The residual strength and modulus of all specimens that achieved run-out were characterized. Composite microstructure, as well as damage and failure mechanisms were investigated.

INTRODUCTION

Advances in power generation systems for aircraft engines, land-based turbines, rockets, and, most recently, hypersonic missiles and flight vehicles have raised the demand for structural materials that have superior long-term mechanical properties and retained properties under high temperature, high pressure, and varying environmental factors, such as moisture[1]. Typical components include combustors, nozzles and thermal insulation. Ceramic-matrix composites (CMCs), capable of maintaining excellent strength and fracture toughness at high temperatures are prime candidate materials for such applications. Additionally, lower densities of CMCs and their higher use temperatures, together with a reduced need for cooling air, allow for improved high-temperature performance when compared to conventional nickel-based superalloys[2]. Advanced reusable space launch vehicles will likely incorporate fiber-reinforced CMCs in critical propulsion components[3]. Because these applications require exposure to oxidizing environments, the thermodynamic stability and oxidation resistance of CMCs are vital issues.

The main advantage of CMCs over monolithic ceramics is their superior toughness, tolerance to the presence of cracks and defects, and non-catastrophic mode of failure. It is widely accepted that in order to avoid brittle fracture behavior in CMCs and improve the damage tolerance, a weak fiber/matrix interface is needed, which serves to deflect matrix cracks and to allow subsequent fiber pullout[4-7]. Historically, following the development of SiC fibers, fiber coatings such as C or BN have been employed to promote the desired composite behavior. However, the non-oxide fiber/non-oxide matrix composites generally show poor oxidation resistance[8,9], particularly at intermediate

---

* Corresponding author

The views expressed are those of the authors and do not reflect the official policy or position of the United States Air Force, Department of Defense or the U. S. Government.

151

temperatures (~800 °C). These systems are susceptible to embrittlement due to oxygen entering through the matrix cracks and then reacting with the interphase and the fibers[10-13]. The degradation, which involves oxidation of fibers and fiber coatings, is typically accelerated by the presence of moisture[14-20]. Using oxide fiber/ non-oxide matrix or non-oxide fiber/oxide matrix composites generally does not substantially improve the high-temperature oxidation resistance[21]. The need for environmentally stable composites motivated the development of CMCs based on environmentally stable oxide constituents[22-30].

More recently it has been demonstrated that similar crack-deflecting behavior can also be achieved by means of a finely distributed porosity in the matrix instead of a separate interface between matrix and fibers[31]. This microstructural design philosophy implicitly accepts the strong fiber/matrix interface. It builds on the experience with porous interlayers as crack deflection paths[32,33] and extends the concept to utilize a porous matrix as a surrogate. The concept has been successfully demonstrated for oxide-oxide composites[22,26,30,34-38]. Resulting oxide/oxide CMCs exhibit damage tolerance combined with inherent oxidation resistance. However, due to the strong bonding between the fiber and matrix, a minimum matrix porosity is needed for this concept to work[39]. An extensive review of the mechanisms and mechanical properties of porous-matrix oxide/oxide CMCs is given elsewhere[40].

Porous-matrix oxide/oxide CMCs exhibit several behavior trends that are distinctly different from those exhibited by traditional non-oxide CMCs with a fiber-matrix interface. For the non-oxide CMCs, fatigue is significantly more damaging than creep. Zawada et al[41] examined the high-temperature mechanical behavior of a porous matrix Nextel610/Aluminosilicate composite. Results revealed excellent fatigue performance at 1000 °C. Conversely, creep lives were short, indicating low creep resistance and limiting the use of that CMC to temperatures below 1000 °C. Ruggles-Wrenn et al[42] showed that Nextel™720/Alumina (N720/A) composite exhibits excellent fatigue resistance in laboratory air at 1200 °C. The fatigue limit (based on a run-out condition of 10^5 cycles) was 170 MPa (88% UTS at 1200 °C). Furthermore, the composite retained 100% of its tensile strength. However, creep loading was found to be considerably more damaging. Creep run-out (defined as 100 h at creep stress) was achieved only at stress levels below 50% UTS. Creep performance at 1200 °C was further degraded by the presence of steam. Mehrman et al[43] demonstrated that introduction of a short hold period at the maximum stress into the fatigue cycle significantly degraded the fatigue performance of N720/A composite at 1200 °C in air. In steam, superposition of a hold time onto a fatigue cycle resulted in an even more dramatic deterioration of fatigue life, reducing it to the much shorter creep life at a given applied stress.

Because creep was shown to be considerably more damaging than cyclic loading to oxide-oxide CMCs with porous matrix[41-43], high-temperature creep resistance remains among the key issues that must be addressed before using these materials in advanced aerospace applications. In addition, the prior studies[42-43] revealed that creep performance of the N720/A composite deteriorates drastically in the presence of steam. The objective of this study is to identify the temperature range where steam environment causes degradation of creep resistance of Nextel™720/Alumina (N720/A), an oxide-oxide CMC with a porous matrix. This effort investigates tensile creep behavior of N720/A at temperatures in the 1000-1200 °C range in air and in steam environment. Results reveal that the degrading effects of steam on creep performance become pronounced at 1100 °C at creep stresses ≥ 65% of the tensile strength.

MATERIAL AND EXPERIMENTAL ARRANGEMENTS

The material studied was Nextel™720/Alumina (N720/A), an oxide-oxide CMC (manufactured by COI Ceramics, San Diego, CA) consisting of a porous alumina matrix reinforced with Nextel™720 fibers. There is no fiber coating. The damage tolerance of N720/A is enabled by the porous matrix. The composite was supplied in a form of 2.8-mm thick plates comprised of 12 0°/90°

woven layers, with a density of ~2.77 g/cm³, a fiber volume of ~44%, and matrix porosity of ~22%. The fiber fabric was infiltrated with the matrix in a sol-gel process. The laminate was dried with a "vacuum bag" technique under low pressure and low temperature, then pressureless sintered[44]. Representative micrograph of the untested material is presented in Fig. 1(a), which shows 0° and 90° fiber tows as well as numerous matrix cracks. In the case of the as-processed material, most are shrinkage cracks formed during processing rather than matrix cracks generated during loading. Porous nature of the matrix is seen in Fig. 1(b).

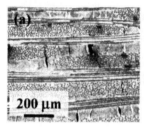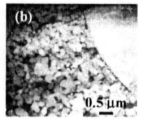

Fig. 1. As-received material: (a) overview, (b) porous nature of the matrix is evident.

A servocontrolled MTS mechanical testing machine equipped with hydraulic water-cooled collet grips, a compact two-zone resistance-heated furnace, and two temperature controllers was used in all tests. An MTS TestStar digital controller was employed for input signal generation and data acquisition. Strain measurement was accomplished with an MTS high-temperature air-cooled uniaxial extensometer of 12.5–mm gage length. For elevated temperature testing, thermocouples were bonded to the specimens using alumina cement (Zircar) to calibrate the furnace on a periodic basis. The furnace controllers (using non-contacting thermocouples exposed to the ambient environment near the test specimen) were adjusted to determine the power settings needed to achieve the desired temperature of the test specimen. The determined power settings were then used in actual tests. Tests in steam environment employed an alumina susceptor (tube with end caps), which fits inside the furnace. The specimen gage section is located inside the susceptor, with the ends of the specimen passing through slots in the susceptor. Steam is introduced into the susceptor (through a feeding tube) in a continuous stream with a slightly positive pressure, expelling the dry air and creating a near 100% steam environment inside the susceptor. The power settings for testing in steam were determined by placing the specimen instrumented with thermocouples in steam environment and repeating the furnace calibration procedure. Thermocouples were not bonded to the test specimens after the furnace was calibrated. Fracture surfaces of failed specimens were examined using an SEM (FEI Quanta 200 HV) as well as an optical microscope (Zeiss Discovery V12). The SEM specimens were carbon coated.

In all elevated-temperature tests, a specimen was heated to test temperature at the rate of 1 °C/s, and held at temperature for additional 15 min prior to testing. Dog bone shaped specimens of 152 mm total length with a 10-mm-wide gage section were used in all tests. Tensile tests were performed in stroke control with a constant displacement rate of 0.05 mm/s in laboratory air. Creep-rupture tests were conducted in load control in accordance with the procedure in ASTM standard C 1337 at 1000, 1100, and 1200 °C in laboratory air and in steam environments. In all creep tests the specimens were loaded to the creep stress level at the stress rate of 15 MPa/s. Creep run-out was defined as 100 h at a given creep stress. In each test, stress-strain data were recorded during the loading to the creep stress level and the actual creep period. Thus both total strain and creep strain could be calculated and examined. To determine the retained tensile strength and modulus, specimens that achieved run-out were subjected tensile test to failure at the temperature of the creep test. It is worthy of note that in all tests reported below, the failure occurred within the gage section of the extensometer.

RESULTS AND DISCUSSION

Monotonic Tension
        The N720/A specimens were tested in tension to failure at 800, 900, 1000, and 1100 °C. Tensile results are summarized in Table I, where elastic modulus, ultimate tensile strength (UTS). and failure strain are presented for each test temperature. The tensile stress-strain curves are shown in Fig. 2. Results from prior work[42] obtained at 23, 1200 and 1330 °C are included in Table I and in Fig. 2 for comparison.

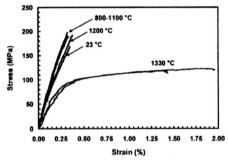

Fig. 2. Tensile stress-strain curves for N720/A ceramic composite at various temperatures. Results at 23, 1200 and 1330 °C from Ruggles-Wrenn et al[42]

        The stress-strain curves obtained in the 23-1200 °C temperature range are nearly linear to failure. Material exhibits typical fiber-dominated composite behavior. Tensile properties change little as the temperature increases from 23 to 1200 °C. The stress-strain behavior changes dramatically at 1330 °C. The stress-strain curve is largely nonlinear. the strength and modulus decrease significantly, while the failure strain increases.

Table I. Summary of tensile properties for the N720/A ceramic composite. Results at 23, 1200 and 1330 °C from Ruggles-Wrenn et al[42].

Temperature (°C)	Elastic Modulus (GPa)	UTS (MPa)	Failure Strain (%)
23 *	60.3	169	0.35
800	69.4	197	0.33
900	72.2	189	0.34
900	68.0	191	0.31
1000	72.6	183	0.31
1000	74.4	192	0.32
1100	70.8	189	0.33
1100	68.4	191	0.32
1200 *	74.7	192	0.38
1330 *	42.6	120	1.68

* Data at 1200 °C from Ruggles-Wrenn et al[42]

Creep-Rupture

Tensile creep tests were conducted at 1000 and 1100 °C in air and in steam. Results are summarized in Table II, where creep strain accumulation and rupture time are shown for each creep stress level, test temperature and environment. Creep-rupture results from prior work[42] obtained at 1200 °C are included in Table II for comparison. Creep curves obtained at 1000 and 1100 °C are shown in Figs. 3 and 4, respectively. Creep curves from prior work[42] obtained at 1200 °C are shown in Fig. 5. Note that at each temperature creep tests were conducted both in air and in steam. The time scale in Fig. 5(b) is reduced to clearly show the creep curves obtained at stress levels ≥ 125 MPa.

Table II. Summary of creep-rupture results for the N720/A ceramic composite at various temperatures in laboratory air and in steam. Results at 1200 °C from Ruggles-Wrenn et al[42].

Temperature (°C)	Creep Stress (MPa)	Creep Strain (%)	Time to Rupture (h)
Test in laboratory air			
1000	150	0.08	100 [a]
1100	150	0.13	100 [a]
1200 [b]	80	0.59	100 [a]
1200 [b]	100	1.52	41.0
1200 [b]	125	1.28	18.1
1200 [b]	150	0.58	0.27
Test in steam			
1000	135	0.13	100 [a]
1000	150	0.14	100 [a]
1000	160	0.18	100 [a]
1100	100	0.50	100 [a]
1100	125	0.49	53.7
1100	150	0.46	12.2
1200 [b]	80	2.96	46.0
1200 [b]	100	1.41	2.49
1200 [b]	125	0.90	0.24
1200 [b]	150	0.40	0.03

[a] Run-out
[b] Data from Ruggles-Wrenn et al[42]

Creep curves produced in all tests conducted in air exhibit primary and secondary creep regimes. At 1000 °C, transition from primary to secondary creep occurs late in creep life, primary creep persists during the first 40-50 h of the creep test. At 1100 and especially at 1200°C, transition from primary to secondary creep occurs earlier. As expected, creep strain accumulation increases with test temperature. Creep strains accumulated in air at 1000 and 1100 °C are considerably lower than the corresponding failure strains obtained in tension tests. Conversely, all creep strains accumulated in air at 1200 °C significantly exceed the failure strain obtained in the tension test at that temperature. Creep run-out is achieved in all tests conducted in air at 1000 and 1100 °C. At 1200 °C in air, creep run-out was achieved only at 80 MPa.

Results in Figs. 3-5 reveal that test environment has little influence on the appearance of the creep curves for the N720/A composite. The creep curves produced at each temperature in steam are qualitatively similar to the creep curves obtained at that temperature in air.

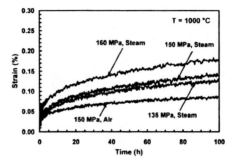

Fig. 3. Creep strain vs time curves for N720/A ceramic composite at 1000 °C in laboratory air and in steam environment.

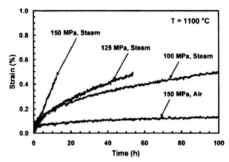

Fig. 4. Creep strain vs time curves for N720/A ceramic composite at 1100 °C in laboratory air and in steam environment.

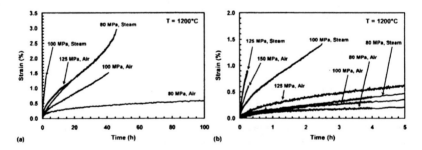

Fig. 5. Creep strain vs time curves for N720/A CMC at 1200 °C in air and in steam: (a) time scale chosen to show creep strains accumulated at 80 MPa and (b) time scale reduced to show creep curves at stresses ≥ 125 MPa. Data from Ruggles-Wrenn et al[42].

The creep lifetimes obtained at 1000 °C in steam are comparable to those produced in air. All tests conducted in steam at 1000 °C achieved a 100-h run-out. However, the steam environment has a noticeable effect on creep strain accumulated at 1000 °C. At 150 MPa, the creep strain accumulated during 100 h of creep in steam is nearly two times that accumulated in air. At 1100 °C the presence of steam has a more pronounced effect on both creep strain and creep lifetime. In steam at 1100 °C, creep run-out was achieved only at 100 MPa. Creep strain produced at 150 MPa at 1100 °C in steam is nearly 3.5 times that obtained in air. At 1200 °C the effects of steam are amplified considerably. While N720/A survived 100 h of creep at 80 MPa at 1200 °C in air, creep run-out was not achieved in steam. Furthermore, creep strain produced at 80 MPa at 1200 °C in steam was 5 times that produced in air.

Minimum creep rate was reached in all tests. Creep rate as a function of applied stress is presented in Fig. 6, where results obtained in air and in steam at 1200 °C from prior work[42] are included for comparison. As expected, the creep strain rates increase with increasing applied stress as well as with increasing temperatures. It is seen that in air the secondary creep rate obtained at 1200 °C can be as high as 10^3 times those produced at 1000 and 1100°C. At 1200 °C in air, the minimum creep rate increases by approximately two orders of magnitude as the creep stress increases from 80 to 150 MPa. At 1000 °C the presence of steam has negligible effect on creep strain rate. Creep rates obtained in all tests conducted at 1000 °C are $\leq 10^{-9}$ s^{-1}. As the temperature increases to 1100 °C, the creep rate obtained in air remains close to that obtained at 1000°C, but the creep rates obtained in steam accelerate dramatically. At 1100 °C at 150 MPa, the creep rate obtained in steam is two orders of magnitude higher than the rate obtained in air. At 1200 °C, creep rates in steam are approximately one order of magnitude higher than the creep rates obtained in air for a given applied stress.

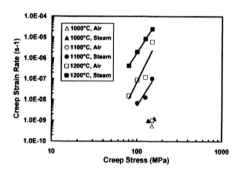

Fig. 6. Minimum creep rate as a function of applied stress for N720/A CMC at various temperatures in air and in steam. Data at 1200 °C from Ruggles-Wrenn et al[42].

Stress-rupture behavior is summarized in Fig. 7, where results obtained at 1200 °C from prior work[42] are also included. In air, an increase in temperature from 1000 to 1100 °C has no effect on creep lifetime up to 100 h. With the creep run-out condition defined as 100 h, 150 MPa is the run-out stress in air at both 1000 and 1100 °C. However, a further temperature increase to 1200 °C significantly degrades creep lifetime. For the applied stress of 150 MPa, creep life at 1200 °C was reduced by 99% compared to that at 1100 °C. At 1200 °C in air, the creep run-out stress was only 80 MPa. At 1000 °C the presence of steam has no influence on creep lifetimes (up to 100 h). In steam at 1000 °C creep run-out was achieved at 160 MPa. At 1100 °C the presence of steam dramatically reduced creep lifetimes. In steam at 1100 °C creep run-out was achieved only at 100 MPa. For the applied stress of 150 MPa, the reduction in creep life due to steam was ≈ 88%. An even greater degradation of creep life due to the presence of steam is seen at 1200 °C. At 1200 °C in steam creep run-out was not achieved. The loss of creep life due to steam at 1200 °C was at least 90% for applied stress levels ≥100 MPa, and 54% for the applied stress of 80 MPa.

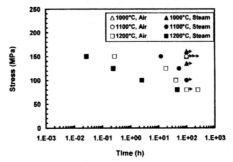

Fig. 7. Creep stress vs time to rupture for N720/A ceramic composite at various temperatures in air and in steam. Data at 1200 °C from Ruggles-Wrenn et al[42].

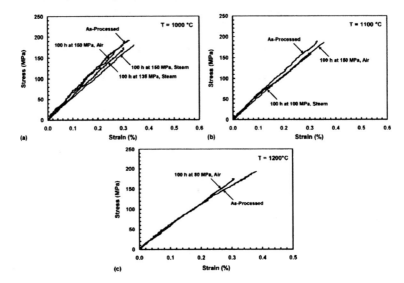

Fig. 8. Effects on tensile stress-strain behavior of prior creep at: (a) 1000 °C, (b) 1100 °C and (c) 1200 °C, data from Ruggles-Wrenn et al[42].

Retained strength and modulus of the specimens that achieved a run-out are summarized in Table III. Tensile stress-strain curves obtained for the specimens subjected to prior creep are presented in Fig. 8 together with the tensile stress-strain curves for the as-processed material. Results from prior work[42] obtained at 1200 °C are included in Table III and in Fig. 8 for comparison. Prior creep in air at 1000 and 1100 °C had little effect on tensile strength. However, a reduction in modulus was observed. The modulus loss was 14% for the specimen pre-crept in air at 1000 °C and 16% for the specimen pre-crept in air at 1100 °C. Conversely, prior creep in steam at 1000 and 1100 °C considerably decreased both tensile strength and stiffness. At 1000 °C strength loss due to prior creep in steam reached 9% and modulus loss, 26%. In the case of the specimen pre-crept in steam at 1100 °C, the strength loss was nearly 16% and modulus loss was ~25%. At 1200 °C, prior creep in air reduced the tensile strength by nearly 8%, and the tensile modulus, by 15%.

Table III. Retained properties of the N720/A specimens subjected to prior creep at 1000, 1100 and 1200 °C in laboratory air and steam environments. Results at 1200 °C from Ruggles-Wrenn et al[42].

Environment	Creep Stress (MPa)	Retained Strength (MPa)	Retained Modulus (GPa)	Strain at Failure (%)
		Test at 1000 °C		
Air	150	180	62.9	0.29
Steam	135	178	62.1	0.34
Steam	150	174	56.9	0.30
Steam	160	170	54.0	0.30
		Test at 1100 °C		
Air	150	186	58.3	0.36
Steam	100	160	52.5	0.30
		Test at 1200 °C[a]		
Air	80	174	64.5	0.31

[a] Data at 1200 °C from Ruggles-Wrenn et al[42]

At 1100 and, especially, at 1200 °C, the presence of steam dramatically reduced creep lifetimes. Because the creep performance of the composite with 0°/90° orientation is dominated by the fibers, fiber degradation is a likely source of the composite degradation. It is recognized that stress corrosion of the N720 fibers may be the mechanism behind reduced creep resistance at 1100 and 1200°C in steam. In this case, subcritical (slow) crack growth in the fiber is caused by a chemical interaction of water molecules with mechanically strained Si-O bonds at the crack tip, with the rate of chemical reaction increasing exponentially with applied stress[45-53]. For glass and ceramic materials which have slow crack growth due to stress corrosion as a unique, time-dependent failure mechanism, it is possible to predict the cyclic fatigue lifetime from the static fatigue (creep) data by using a linear elastic crack growth model[54]. Ruggles-Wrenn et al[55] applied the fracture mechanics approach proposed by Evans and Fuller[54] to the cyclic and static fatigue data obtained for N720/A at 1200 °C in steam. Excellent agreement between predicted cyclic lifetimes and experimental results in steam showed that slow crack growth due to stress corrosion was indeed the governing failure mechanism at 1200 °C in steam. The results of the present study suggest that in steam the time-dependent failure mechanism proceeds at a slow rate at 1000 °C, but accelerates significantly as the temperature increases to 1200 °C.

Composite Microstructure

Optical micrographs of fracture surfaces obtained in the 150 MPa creep tests conducted in air and in steam at 1000, 1100 and 1200°C are shown in Fig. 9. In general the fracture surfaces of the N720/A specimens tested in air (Figs. 9 (a)-(c)) exhibit larger amounts of fiber pullout and longer damage zones than those of the N720/A specimens tested in steam (Figs. 9 (d)-(f)). This difference is particularly striking in the case of specimens tested at 1200 °C. The N720/A fracture surface obtained at 1200 °C in air (Fig. 9(c)) shows uncoordinated fiber failure, while the fracture surface obtained in steam (Fig. 9(f)) exhibits hardly any fibrous fracture. It is seen that the N720/A specimens tested at

1000 °C in air and in steam (Figs. 9 (a) and (d)) and the N720/A specimen tested at 1100 °C in air (Fig. 9(b)), which achieved 100-h run-out at 150 MPa and failed in a subsequent tension tests, have considerably longer damage zones than the specimen tested at 1100 °C in steam (Fig. 9(e)) and the specimens tested at 1200 °C (Figs. 9 (c) and (f)). The N720/A specimens tested in this effort produced damage zones ranging up to 15 mm in length. It is noteworthy that specimens that exhibited longer lifetimes invariably produced longer damage zones.

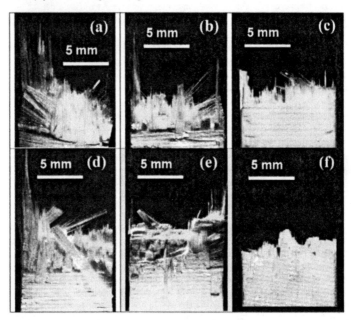

Fig. 9. Fracture surfaces obtained in creep tests at 150 MPa: (a) at 1000 °C in air, (b) at 1100 °C in air, (c) at 1200 °C in air, (d) at 1000 °C in steam, (e) at 1100 °C in steam, (f) at 1200 °C in steam.

Further understanding of the influence of temperature and environment on the fracture surface topography and the microstructure of N720/A specimens tested in creep at 150 MPa in air and in steam can be gained by examining SEM micrographs in Figs. 10 and 11. The fracture surface obtained at 1000 °C in air (Fig. 10(a)) is dominated by the regions of uncorrelated fiber fracture, where individual fibers are clearly discernable. Note that the SEM micrograph in Fig. 10(b) reveals only small amounts of matrix particles bonded to the fiber surfaces. In contrast, the fracture surface produced at 1200 °C (Fig. 10(c)) is dominated by planar regions of coordinated fiber failure. As seen in the SEM micrograph in Fig. 10(d), the amount of matrix material remaining bonded to the fibers is significantly greater than that in the specimen subjected to creep at 1000 °C.

The test environment has a dramatic effect on the fracture surface appearance. While the fracture surface produced at 1000 °C in steam (Fig. 11 (a)) still exhibits some areas of uncoordinated brushy failure, considerable areas of planar fracture are also visible. The SEM micrograph in Fig. 11(b) reveals that a somewhat greater amount of matrix material remains bonded to the pulled-out fibers in a specimen tested at 1000 °C in steam compared to that tested at 1000 °C in air. Whereas the fracture surface obtained at 1200 °C in air (Fig. 11(c)) still shows some limited areas of fibrous fracture, the fracture surface produced at 1200 °C in steam (Fig. 11(c)) is almost entirely planar. In the case of the specimen tested at 1200 °C in steam (Fig. 11(d)), the amounts of matrix material adhering to the fibers are much greater than those seen in other specimens.

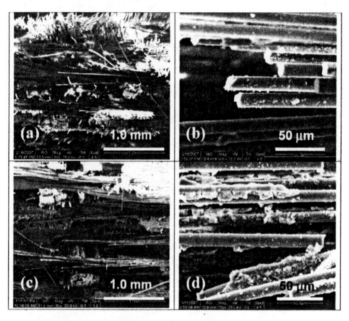

Fig. 10. Fracture surfaces of the N720/A specimens tested in creep at 150 MPa in air: (a)-(b) at 1000 °C, and (c)-(d) at 1200 °C.

Notably, as the temperature increases, the extent of correlated fiber failure increases and the creep lifetime decreases. It is recognized that the increase in the spatial correlation in the fiber failure locations is among the main manifestations of the matrix densification[56, 57]. The near-planar fracture surfaces obtained at 1200 °C in air and in steam indicate the loss of matrix porosity and subsequent matrix densification due to additional sintering. As a result, the N720/A composite exhibits decreased damage tolerance and reduced creep lifetimes at 1200 °C. Recall that the fracture surfaces shown in Figs. 10(a) and 11(a) were obtained in tensile tests of the N720/A specimens subjected to 100 h of prior creep at 150 MPa at 1000 °C in air and in steam, respectively. Whereas creep run-out was achieved in both environments, the retained properties of the two specimens were somewhat different. The fracture surface of the N720/A run-out specimen tested in air (Fig. 10(a)) is brushy. This suggests that matrix changes and, consequently, loss of toughness and degradation of retained properties were limited. Recall that the N720/A specimen subjected to 100-h of prior creep at 150 MPa in air retained nearly 100% of its tensile strength. In contrast, the fracture surface of the N720/A run-out specimen tested in steam (Fig. 11(a)) shows fairly extensive areas of coordinated fiber failure and increased fiber-matrix bonding. Apparently 100-h exposure under load at 1000 °C in steam causes changes in the matrix, akin to those observed after shorter exposures under sustained loading at 1200 °C. These changes in the alumina matrix lead to loss of damage tolerance and result in lower retained strength. It is recognized[56-58] that the additional sintering of the matrix is accelerated at higher temperatures. It is possible that the sintering of the matrix is also accelerated in the presence of steam.

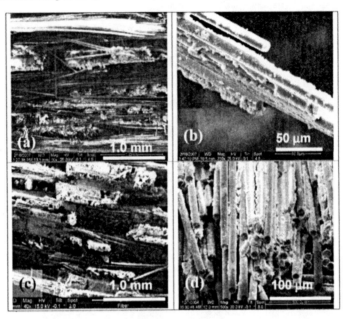

Fig. 11. Fracture surfaces of the N720/A specimens tested in creep at 150 MPa in steam: (a)-(b) at 1000 °C, and (c)-(d) at 1200 °C.

CONCLUDING REMARKS

The creep-rupture behavior of the N720/A composite was characterized in air and in steam for creep stress levels ranging from 135 to 160 MPa at 1000°C, and for stress levels ranging from 100 to 150 MPa at 1100 °C. In air at all temperatures investigated the N720/A composite exhibits primary and secondary creep regimes. At 1000 and 1100 °C, all creep strains accumulated in air are lower than the corresponding failure strains obtained in tension tests.

The test environment has little influence on the appearance of the creep curves of the N720/A composite. At each temperature, the creep curves produced in steam are qualitatively similar to the creep curves obtained in air. At 1000 and 1100 °C, the creep strains accumulated at 150 MPa in steam are significantly larger than those produced in air.

Minimum creep rate was reached in all tests. In air, creep strain rates were below 10^{-9} s^{-1} at 1000 and 1100 °C, and ranged from 1.5×10^{-8} to 6.0×10^{-6} s^{-1} at 1200 °C. The presence of steam has negligible effect on creep strain rates at 1000 °C. The presence of steam accelerates creep rates of N720/A by two orders of magnitude at 1100 °C.

In air, creep run-out was achieved at 150 MPa at both 1000 and 1100 °C. At 1000 °C, the presence of steam had no effect on creep lifetime up to 100 h. At 1100 °C, the presence of steam dramatically reduced creep lifetimes. At 1100 °C in steam creep run-out stress was 100 MPa. At 150 MPa the reduction in creep life due to steam was ≈ 88%.

The N720/A fracture surfaces obtained at 1000 and 1100 °C exhibit regions of uncoordinated brushy failure as well as areas of nearly planar fracture. The fracture surfaces produced at 1000 °C are dominated by regions of fibrous fracture. At 1100 °C, uncorrelated fiber fracture is prevalent in air, and coordinated fiber failure is predominant in steam. The fracture surface appearance can be correlated with failure time. Predominantly planar fracture surface corresponds to a short life, while fibrous fracture indicates longer life. Stress corrosion of the N720 fibers is likely the mechanism behind poor creep resistance of the N720/A composite in steam.

REFERENCES

[1]F. Zok, "Developments in Oxide Fiber Composites," *J. Am. Ceram. Soc.*, **89**(11), 3309-3324 (2006).

[2]L. P. Zawada, J. Staehler, S. Steel, "Consequence of Intermittent Exposure to Moisture and Salt Fog on the High-Temperature Fatigue Durability of Several Ceramic-Matrix Composites," *J. Am. Ceram. Soc.*, **86**(8), 1282-1291 (2003).

[3]S. Schmidt , S. Beyer, H. Knabe, H. Immich, R. Meistring, A. Gessler, "Advanced Ceramic Matrix Composite Materials for Current and Future Propulsion Technology Applications," *Acta Astronautica*, **55**, 409-420 (2004).

[4]R. J. Kerans, R. S. Hay, N. J. Pagano, T. A. Parthasarathy, "The Role of the Fiber-Matrix Interface in Ceramic Composites," *Am. Ceram. Soc. Bull.*, **68**(2), 429-442 (1989).

[5]A. G. Evans, F. W. Zok, "Review: the Physics and Mechanics of Fiber-Reinforced Brittle Matrix Composites," *J. Mater. Sci.*, **29**:3857-3896 (1994).

[6]R. J. Kerans, T. A. Parthasarathy, "Crack Deflection in Ceramic Composites and Fiber Coating Design Criteria," *Composites: Part A*, **30**, 521-524 (1999).

[7]R. J. Kerans, R. S. Hay, T. A. Parthasarathy, M. K. Cinibulk, "Interface Design for Oxidation-Resistant Ceramic Composites," *J. Am. Ceram. Soc.*, **85**(11), 2599- 2632 (2002).

[8]K. M. Prewo, J. A. Batt, "The Oxidative Stability of Carbon Fibre Reinforced Glass-Matrix Composites," *J. Mater. Sci.*, **23**, 523-527 (1988).

[9]T. Mah, N. L. Hecht, D. E. McCullum, J. R. Hoenigman, H. M. Kim, A. P. Katz, H. A. Lipsitt, "Thermal Stability of SiC Fibres (Nicalon)," *J. Mater. Sci.*, **19**, 1191-1201 (1984).

[10]J. J. Brennan. *Fiber Reinforced Ceramic Composites; Ch. 8.* K.C. Masdayazni, editor. Noyes, New York, 1990.

[11]F. Heredia, J. McNulty, F. Zok, A. G. Evans, "Oxidation Embrittlement Probe for Ceramic Matrix Composites," *J. Am. Ceram. Soc.*, **78**(8), 2097-2100 (1995).

[12]R. S. Nutt, "Environmental Effects on High-Temperature Mechanical Behavior of Ceramic Matrix Composites," *High-Temperature Mechanical Behavior of Ceramic Composites.* S. V. Nair, and K. Jakus, editors. Butterworth-Heineman, Boston, MA, (1995).

[13]A. G. Evans, F. W. Zok, R. M. McMeeking, Z. Z. Du, "Models of High-Temperature Environmentally-Assisted Embrittlement in Ceramic Matrix Composites," *J. Am. Ceram. Soc.*, **79**, 2345-52 (1996).

[14]K. L. More, P. F. Tortorelli, M. K. Ferber, J. R. Keiser, "Observations of Accelerated Silicon Carbide Recession by Oxidation at High Water-Vapor Pressures," *J. Am. Ceram. Soc.*, **83**(1), 11-213 (2000).

[15]K. L. More, P. F. Tortorelli, M. K. Ferber, L. R. Walker, J. R. Keiser, W. D. Brentnall, N. Miralya, J. B. Price, "Exposure of Ceramic and Ceramic-Matrix Composites in Simulated and Actual Combustor Environments," *Proceedings of International Gas Turbine and Aerospace Congress*, Paper No. 99-GT-292 (1999).

[16]M. K. Ferber, H. T. Lin, J. R. Keiser, "Oxidation Behavior of Non-Oxide Ceramics in a High-Pressure, High-Temperature Steam Environment," *Mechanical, Thermal, and Environmental Testing and Performance of Ceramic Composites and Components*. M. G. Jenkins, E. Lara-Curzio, and S. T. Gonczy, editors. ASTM STP 1392, 210-215 (2000).

[17]J. A. Haynes, M. J. Lance, K. M. Cooley, M. K. Ferber, R. A. Lowden, D. P. Stinton, "CVD Mullite Coatings in High-Temperature, High-Pressure Air-H_2O," *J. Am. Ceram. Soc.*, **83**(3), 657-659 (2000).

[18]E. J. Opila, R. E. Hann Jr., "Paralinear Oxidation of SiC in Water Vapor," *J. Am. Ceram. Soc.*, **80**(1), 197-205 (1997).

[19]E. J. Opila, "Oxidation Kinetics of Chemically Vapor Deposited Silicon Carbide in Wet Oxygen," *J. Am. Ceram. Soc.*, **77**(3), 730-736 (1994).

[20]E. J. Opila, "Variation of the Oxidation Rate of Silicon Carbide with Water Vapor Pressure," *J. Am. Ceram. Soc.*, **82**(3), 625-636 (1999).

[21]E. E. Hermes, R. J. Kerans, "Degradation of Non-Oxide Reinforcement and Oxide Matrix Composites," *Mat. Res. Soc., Symposium Proceedings*, **125**, 73-78 (1988).

[22]A. Szweda, M. L. Millard, M. G. Harrison, *Fiber-Reinforced Ceramic-Matrix Composite Member and Method for Making*, U. S. Pat. No. 5 601 674, (1997).

[23]S. M. Sim, R. J. Kerans, "Slurry Infiltration and 3-D Woven Composites," *Ceram. Eng. Sci. Proc.*, **13**(9-10), 632-641 (1992).

[24]E. H. Moore, T. Mah, and K. A. Keller, "3D Composite Fabrication Through Matrix Slurry Pressure Infiltration," *Ceram. Eng. Sci. Proc.*, **15**(4), 113-120 (1994).

[25]M. H. Lewis, M. G. Cain, P. Doleman, A. G. Razzell, J. Gent, "Development of Interfaces in Oxide and Silicate Matrix Composites," *High-Temperature Ceramic–Matrix Composites II: Manufacturing and Materials Development*, A. G. Evans, and R. G. Naslain, editors, American Ceramic Society, 41–52 (1995).

[26]F. F. Lange, W. C. Tu, A. G. Evans, "Processing of Damage-Tolerant, Oxidation-Resistant Ceramic Matrix Composites by a Precursor Infiltration and Pyrolysis Method," *Mater. Sci. Eng. A*, **A195**, 145–150 (1995).

[27]R. Lunderberg, L. Eckerbom, "Design and Processing of All-Oxide Composites," *High-Temperature Ceramic–Matrix Composites II: Manufacturing and Materials Development*, A. G. Evans, and R. G. Naslain, editors, American Ceramic Society, 95–104 (1995).

[28]E. Mouchon, P. Colomban, "Oxide Ceramic Matrix/Oxide Fiber Woven Fabric Composites Exhibiting Dissipative Fracture Behavior," *Composites*, **26**, 175–182 (1995).

[29]P. E. D. Morgan and D. B. Marshall, "Ceramic Composites of Monazite and Alumina," *J. Am. Ceram. Soc.*, **78**(6), 1553–1563 (1995).

[30]W. C. Tu, F. F. Lange, A. G. Evans, "Concept for a Damage-Tolerant Ceramic Composite with Strong Interfaces," *J. Am. Ceram. Soc.*, **79**(2), 417–424 (1996).

[31]C. G. Levi, J. Y. Yang, B. J. Dalgleish, F. W. Zok, A. G. Evans, "Processing and Performance of an All-Oxide Ceramic Composite," *J. Am. Ceram. Soc.*, **81**, 2077-2086 (1998).

[32]J. B. Davis, J. P. A. Lofvander, A. G. Evans, "Fiber Coating Concepts for Brittle Matrix Composites," *J. Am. Ceram. Soc.*, **76**(5), 1249–57 (1993).

[33]T. J. Mackin, J. Y. Yang, C. G. Levi, A. G. Evans, "Environmentally Compatible Double Coating Concepts for Sapphire Fiber Reinforced γ-TiAl," *Mater. Sci. Eng.*, **A161**, 285–93 (1993).

[34]A. G. Hegedus, *Ceramic Bodies of Controlled Porosity and Process for Making Same*, U. S. Pat. No. 5 0177 522, May 21, (1991).

[35]T. J. Dunyak, D. R. Chang, M. L. Millard, "Thermal Aging Effects on Oxide/Oxide Ceramic-Matrix Composites," *Proceedings of 17th Conference on Metal Matrix, Carbon, and Ceramic Matrix Composites. NASA Conference Publication 3235, Part 2*, 675-90 (1993).

[36]L. P. Zawada, S. S. Lee, "Mechanical Behavior of CMCs for Flaps and Seals," *ARPA Ceramic Technology Insertion Program (DARPA), W. S. Coblenz WS, editor.* Annapolis MD, 267-322 (1994).

[37]L. P. Zawada, S. S. Lee, "Evaluation of the Fatigue Performance of Five CMCs for Aerospace Applications," *Proceedings of the Sixth International Fatigue Congress*, 1669-1674 (1996).

[38]T. J. Lu, "Crack Branching in All-Oxide Ceramic Composites," *J. Am. Ceram. Soc.*, **79**(1), 266-274 (1996).

[39]M.A. Mattoni, J.Y. Yang, C.G. Levi, F.W. Zok, "Effects of Matrix Porosity on the Mechanical Properties of a Porous Matrix, All-Oxide Ceramic Composite," *J. Am. Ceram. Soc.*, **84**(11), 2594-2602 (2003).

[40]F. W. Zok, C. G. Levi, "Mechanical Properties of Porous-Matrix Ceramic Composites," *Adv. Eng. Mater.*, **3**(1-2), 15-23 (2001).

[41]L. P. Zawada, R. S. Hay, S. S. Lee, J. Staehler, "Characterization and High-Temperature Mechanical Behavior of an Oxide/Oxide Composite," *J. Am. Ceram. Soc.*, **86**(6), 981-90 (2003).

[42]M. B. Ruggles-Wrenn, S. Mall, C. A. Eber, L. B. Harlan, "Effects of Steam Environment on High-Temperature Mechanical Behavior of Nextel[TM]720/Alumina (N720/A) Continuous Fiber Ceramic Composite," *Composites: Part A*, **37**(11), 2029-40 (2006).

[43]J. M. Mehrman, M. B. Ruggles-Wrenn, S. S. Baek, "Influence of Hold Times on the Elevated-Temperature Fatigue Behavior of an Oxide-Oxide Ceramic Composite in Air and in Steam Environment," *Comp. Sci. Tech.*, **67**, 1425-1438 (2007).

[44]R. A. Jurf, S. C. Butner, "Advances in Oxide-Oxide CMC," *J. Eng. Gas. Turbines Power, Trans ASME*, **122**(2), 202-205 (1999).

[45]R. J. Charles and W. B. Hillig WB, "The Kinetics of glass Failure by Stress Corrosion, *Symposium on Mechanical Strength of Glass and Ways of Improving It.* Florence, Italy. September 25-29 (1961). Union Scientifique Continentale du Verre, Charleroi, Belgium, 511-27 (1962).

[46]R. J. Charles and W. B. Hillig WB, "Surfaces, Stress-Dependent Surface Reactions, and Strength," *High-Strength Materials.* V. F. Zackey, editor. John Wiley & Sons, Inc., New York, 682-705 (1965).

[47]S. M. Wiederhorn, "Influence of Water Vapor on Crack Propagation in Soda-Lime Glass," *J. Am. Ceram. Soc.*, **50**(8), 407-14 (1967).

[48]S. M. Wiederhorn, L. H. Bolz, "Stress Corrosion and Static Fatigue of Glass," *J. Am. Ceram. Soc.*, **53**(10), 543-48 (1970).

[49]S. M. Wiederhorn, "A Chemical Interpretation of Static Fatigue," *J. Am. Ceram. Soc.*, **55**(2):81-85 (1972).

[50]S. M. Wiederhorn, S. W. Freiman, E. R. Fuller, C. J. Simmons, "Effects of Water and Other Dielectrics on Crack Growth," *J. Matl. Sci.*, **17**,3460-78 (1982).

[51]M T. A. Michalske, S. W. Freiman, "A Molecular Mechanism for Stress Corrosion in Vitreous Silica," *J. Am. Ceram. Soc.*, **66**(4):284-288 (1983).

[52]T. A. Michalske, B. C. Bunker, "Slow Fracture Model Based on Strained Silicate Structures. J Appl. Phys., **56**(10), 2686-93 (1984).

[53]T. A. Michalske, B. C. Bunker, "A Chemical Kinetics Model for Glass Fracture," *J. Am. Ceram. Soc.*, **76**(10), 2613-18 (1993).

[54]A. G. Evans, E. R. Fuller, "Crack Propagation in Ceramic Materials under Cyclic Loading Conditions," *Metall. Trans.*, **5**A(1), 27-33 (1974).

[55]M. B. Ruggles-Wrenn, G. Hetrick, S. S. Baek, "Effects of Frequency and Environment on Fatigue Behavior of an Oxide-Oxide Ceramic Composite at 1200 °C," *Int. J. Fatigue*, **30**, 502-516 (2008).

[56]H. Fujita, G. Jefferson, R. M. McMeeking, F. W. Zok, "Mullite/Alumina Mixtures for Use as Porous Matrices in Oxide Fiber Composites," *J. Am. Ceram. Soc.*, **87**(2), 261-67 (2004).

[57]H. Fujita, C. G. Levi, F. W. Zok, G. Jefferson, "Controlling Mechanical Properties of Porous Mullite/Alumina Mixtures via Precursor-Derived Alumina," *J. Am. Ceram. Soc.*, **88**(2), 367-75 (2005).
[58]E. A. V. Carelli, H. Fujita, J. Y. Yang, F. W. Zok, "Effects of Thermal Aging on the Mechanical Properties of a Porous-Matrix Ceramic Composite," *J. Am. Ceram. Soc.*, **85**(3), 595-602 (2002).

CHARACTERIZATION OF FOREIGN OBJECT DAMAGE IN AN OXIDE/OXIDE COMPOSITE AT AMBIENT TEMPERATURE

Sung R. Choi[†] and Donald J. Alexander
Naval Air Systems Command, Patuxent River, MD 20670

ABSTRACT

Foreign object damage (FOD) behavior of an oxide/oxide (N720/AS) ceramic matrix composite (CMC) was determined at ambient temperature using impact velocities ranging from 100 to 400 m/s by 1.59-mm diameter steel ball projectiles. Two different support configurations of target specimens were used: fully supported and partially supported. For the fully supported configuration, frontal contact stress played a major role in generating composite damage, while for the partially supported case both frontal contact and backside flexure stresses were the combined sources of damage generation. The material beneath impact site was compacted in conjunction with the formation of cone cracks. Unlike monolithic silicon nitrides, the oxide/oxide composite did not show any catastrophic failure even at the highest impact velocity of 400 m/s, showing damage tolerance greater in the composite than in silicon nitrides.

INTRODUCTION

Ceramics, because of their brittleness, are susceptible to localized surface damage and/or cracking when subjected to impact by foreign objects. Foreign object damage (FOD) needs to be considered when ceramic materials are designed for structural applications particularly in aeroengines. A considerable amount of work on impact damage of brittle materials by sharp particles as well as by "blunt" particles or by plates has been done both experimentally and analytically, including the assessments of FOD for turbine engine applications [1-14].

In the previous studies [15,16], FOD behavior of two gas-turbine grade monolithic silicon nitrides (AS800 and SN282) was determined. Ceramic target specimens were impacted at their centers by 1.59-mm (diameter) steel ball projectiles in a velocity range from 220 to 440 m/s. The key material parameter on FOD was fracture toughness of a target material. Multiple flaw/damage systems were identified, depending on impact velocity. Effect of projectile materials on FOD was also determined in AS800 using four different projectiles of hardened steel, annealed steel, silicon nitride, and brass balls [17]. For a given target material/impact condition, the hardness of projectile materials was found to be the most important property to control impact damage. The work was extended to a gas-turbine grade, melt-infiltrated (MI) Sylramic™ SiC/SiC ceramic matrix composite (CMC) [18]. Unlike the silicon nitride ceramics, the SiC/SiC composite specimens exhibited no catastrophic failure up to 400 m/s and showed greater resistance to FOD than silicon nitride counterparts.

The current work includes FOD behavior of an N720™ oxide fiber-reinforced aluminosilicate matrix CMC (N720/AS). Oxide/oxide target specimens in a flexure bar configuration were impacted at velocities from 100 to 400 m/s by 1.59-mm-diameter steel ball projectiles. The main purpose of this paper was to characterize damage morphologies and related flaw/crack systems of the target oxide/oxide composite. Information regarding post-impact strengths and their associated aspects can be found in a separate paper [19].

[†] Corresponding author, sung.choi1@navy.mil

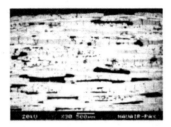

Figure 1. Microstructure of oxide/oxide (N720/AS) ceramic matrix composite used in this work.

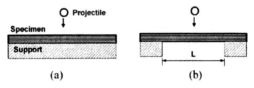

Figure 2. Two types of target specimen support used in FOD testing: (a) fully supported and (b) partially supported (L=20 mm).

EXPERIMENTAL PROCEDURES

Material
        The material used in this work was 2-D woven, N720™ fiber-reinforced aluminosilicate matrix CMCs. N720™ oxide fibers, produced in tow form by 3M Corp. (Minneapolis, MN), were woven into 2-D 8 harness-satin cloth. The cloth was cut into a proper size, slurry-infiltrated with the matrix, and 12 ply-stacked followed by consolidation and sintering. No interface fiber coating was employed. The fiber volume fraction of the composite panels was about 0.45. Typical microstructure of the composite is shown in Fig. 1. Significant porosity and microcracks in the matrix were typical of the composite, intended to improve damage tolerance [20,21]. Porosity was about 20-25 %, bulk density was 2.55 g/cm^3, and elastic modulus was 67 GPa by the impulse excitation of vibration technique. Flexure bars measuring 10 mm in width, 50 mm in length, and about 3 mm in as-furnished thickness were machined from the composite panels for FOD testing.

Foreign Object Damage Testing
        Foreign object damage testing was conducted using the experimental apparatus whose descriptions can be found elsewhere [15,16]. Briefly, hardened (HRC≥60) chrome steel-balls with a diameter of 1.59 mm were inserted into a 300mm-long gun barrel. A helium-gas cylinder and relief valves were utilized to pressurize a reservoir to a specific level, depending on prescribed impact velocity. Upon reaching a specific level of pressure, a solenoid valve was instantaneously opened accelerating a steel-ball projectile through the gun barrel to impact a target specimen. The target specimens were fully or partially supported through a SiC specimen holder, as shown in Fig. 2. Each target specimen was aligned such that the projectile impacted at

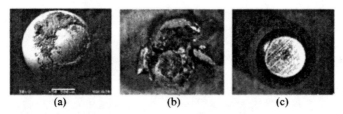

(a)   (b)   (c)

Figure 3. Steel ball projectiles after impact on: (a) oxide/oxide at 400 m/s, (b) AS800 silicon nitride at 400 m/s [17], and (C) AS800 silicon nitride at 200 m/s [17].

the center of the specimen with a normal incidence angle. The range of impact velocity employed was from 100 to 400 m/s. Impact morphologies of target specimens and some recovered projectiles were scrutinized after impact using optical and scanning electron microscopy.

RESULTS AND DISCUSSION

It has been observed that steel-ball projectiles impacting monolithic silicon nitrides (AS800 and SN282) and MI SiC/SiC composite were flattened or severely deformed or fragmented, depending on impact velocity [15-18]. However, the projectiles that impacted the oxide/oxide composite were not flattened or noticeably deformed even at the highest impact velocity of 400 m/s, as shown in Fig. 3. This is attributed to the composite's 'soft' and open (see also Fig. 1) structure, which also can be known by comparing elastic modulus of each material: 67, 340, and 220 GPa, respectively, for the oxide/oxide, the silicon nitrides, and the MI SiC/SiC. Also included in the figure are projectiles having impacted AS800 silicon nitride, showing complete fragmentation at 400 m/s and significant deformation even at 200 m/s.

Front impact damages generated in target specimens were typically in the form of craters. Figure 4 shows impact sites on target specimens in both fully and partially supported configurations. At a low impact velocity of 100 m/s, impact site was not easily observable (Note that the composite was 'white' in color). When impact velocity was above 200 m/s, impact sites were well developed with their size being dependent on impact velocity. Damage size as a function of impact velocity is shown in Fig. 5. The damage size increased *monotonically* with increasing impact velocity and was independent of the type of specimen support. This is again indicative of the oxide/oxide's *soft* and open structure. By contrast, the *hard* and dense MI SiC/SiC composite exhibited significant rate of increase in damage size with impact velocity, as seen in the figure. The difference in damage in MI SiC/SiC between two types of support was found to be insignificant [18].

Backside damages in the target specimens were sensitive to the type of specimens support. At lower impact velocities ≤220 m/s, the backside damage appeared to be indistinguishable between the two supports. However, a remarkable difference in backside damage started at impact velocity ≥300 m/s: No damage was observable in full support; whereas,

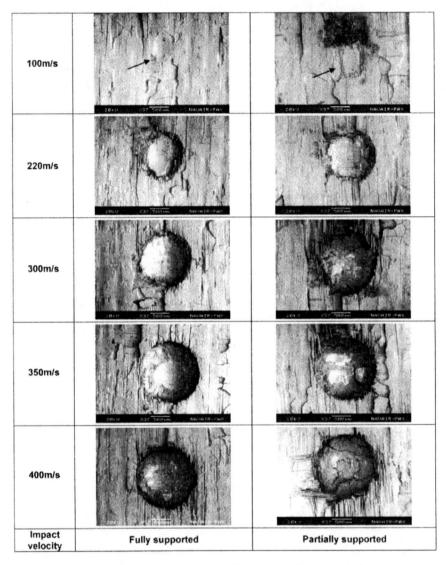

Figure 4. Summary of front impact damage with respect to impact velocity in oxide/oxide composite impacted by 1.59-mm steel ball projectiles in two types of full and partial specimen supports. Arrows indicate impact sites.

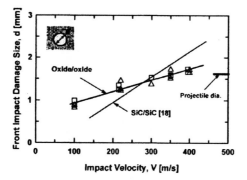

Figure 5. Front impact-damage size as a function of impact velocity for the oxide/oxide composite in full and partial supports. The data on the MI SiC/SiC composite [18] was included for comparison. Δ: full support; ⊔: partial support.

appreciable damage took place in partial support, due to the presence of backside flexural tensile stress. Comparison in backside damage between the two different specimen supports is presented in Fig. 6. At 400m/s, the target specimen in partial support was on the verge of being penetrated by the impacting projectile.

Figure 7 shows cross-sectional views of impact sites with respect to impact velocity in both types of specimen support. In the full support configuration, impact site at 100 m/s was barely developed with a slight contour of deformation. With increasing impact velocity, front impact damage increased, as expected. Of a special importance was the aspect of the material beneath the impact which showed material's compaction (densification). This also can be seen in the figure where the area beneath the impact does not show any horizontal pores. This compaction can be understandable by considering that the material is *soft* and *open* in its structure. Also, note that no backside damage was observable even at 400 m/s, as aforementioned in Fig. 6. It is certain that compaction increased with increasing impact velocity, despite no quantitative evidence presented in this paper. A similar aspect of compaction also occurred in the partial support configuration as seen in the figure. However, backside damage became predominant with increasing impact velocity and therefore significantly differentiated in its degree from that in the full support counterpart (see also Fig. 6). Finally, it should be noted that the oxide/oxide target specimens in either full or partial support survived impact even at 400 m/s. This is in contrast with AS800 and SN282 silicon nitrides that exhibited catastrophic fracture upon impact at much lower velocities of 180-260 m/s.

Of another importance from the cross-sectional views of impact sites was the formation of cone cracks initiating from the impact site, making damage more significant at the backside than at the front impact site and letting the material beneath impact site be displaced toward backside. Generation of cone cracks by spherical projectiles has been observed in monolithic silicon nitrides [4,9,15-17,22] and MI SiC/SiC composite and is typified of many brittle solids including glass dynamically [1,2,5] or statically [8]. Examples of MI SiC/SiC and AS800 silicon

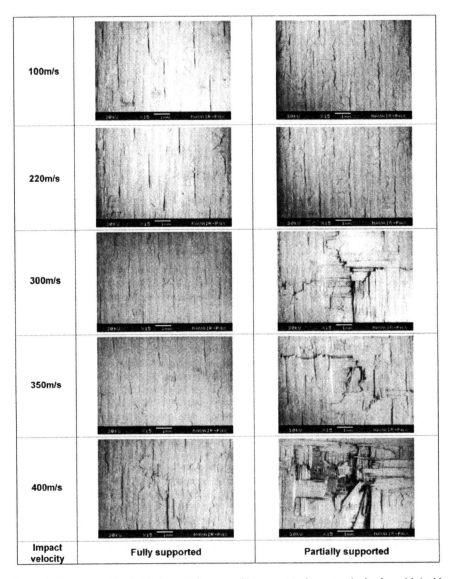

Figure 6. Summary of backside impact damage with respect to impact velocity in oxide/oxide composite impacted by 1.59-mm steel ball projectiles in two types of full and partial specimen supports.

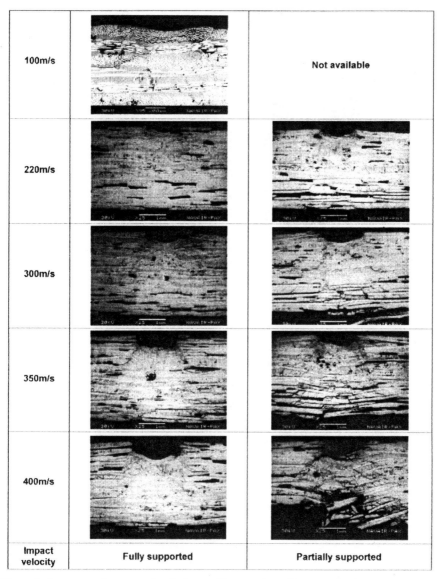

Figure 7. Summary of cross-sectional views of impact sites with respect to impact velocity in oxide/oxide composite impacted by 1.59-mm steel ball projectiles in two types of full and partial specimen supports.

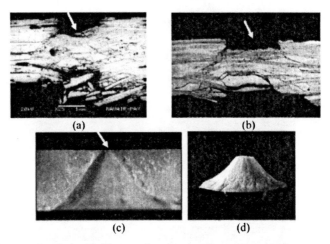

(a)   (b)

(c)   (d)

Figure 8. Cross-sectional views of impact sites showing the occurrence of cone cracking at 400 m/s: (a) oxide/oxide (partial support), (b) MI SiC/SiC (partial support), (c) AS800 silicon nitride (full support) [16], and (d) a cone separated from (c). Arrows indicate impact sites.

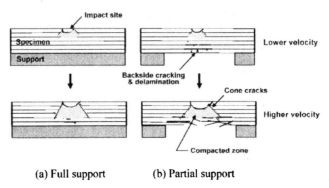

(a) Full support   (b) Partial support

Figure 9. Progression of impact damage with velocity in (a) fully supported and (b) partially supported specimens. The vertical arrow indicates a case of increasing impact velocity.

nitride showing cone cracking are presented and compared with the oxide/oxide composite in Fig. 8. Note that AS800 silicon nitride revealed the contour of a well-developed cone crack from its fracture surface, together with a *volcanic-island* like cone completely separated from the specimen [16]. AS800 and SiC/SiC composite specimens were in 2mm-thick and subjected to

1.59 mm steel-ball impact at 400 m/s in full (for AS800) or partial (SiC/SiC) support. The cone angle was about 80° and 120°, respectively, for AS800 and SiC/SiC composite.

Progressive impact damage with increasing impact velocity for both fully supported and partially supported specimens are illustrated in Fig. 9. It is again noted that regardless of the type of support, cone cracking was one of the dominant impact damage mechanisms and that peculiar to the oxide/oxide composite was a phenomenon of material's compaction beneath the impact site. Contrast to the case for monolithic ceramics, conventional optical and/or SEM examinations for the composite may not be sufficient in many cases. Additional use of pertinent nondestructive evaluation tools such as computed tomography and/or pulsed thermography techniques [23] is strongly recommended. The morphological approaches described above can give qualitative information pertaining to impact damage. However, they do not provide quantitative assessments as to how impact damage results in strength degradation. This can be done by determining post-impact strength of target specimens impacted. Results on post-impact strength testing showed that strength was a decreasing function of impact velocity and that strength degradation was greater in partial support than in full support. Detailed information on strength degradation of the oxide/oxide composite can be found in a separate publication [19].

CONCLUSIONS

1) The overall impact damage of the oxide/oxide composite was found to be greater in partially supported specimens than for fully supported ones.
2) For fully supported specimens, frontal contact stresses together with cone cracking played a major role in generating composite damage; whereas, for partially supported ones, both frontal contact and backside flexure stresses were combined sources of damage generation.
3) Due to the composite's *soft* and *open* structure, compaction of the material beneath impact site occurred with its degree being dependent on impact velocity. Also, the formation of cone cracks took place making damage more significant at backside than at impact site.

Acknowledgements
This work was supported by the Office of Naval Research.

REFERENCES
1. Wiederhorn, S. M., and Lawn, B.R., 1977, "Strength Degradation of Glass Resulting from Impact with Spheres," J. Am. Ceram. Soc., 60[9-10], pp. 451-458.
2. Wiederhorn, S. M., and Lawn B. T., 1979, "Strength Degradation of Glass Impact with Sharp Particles: I, Annealed Surfaces," J. Am. Ceram. Soc., 62[1-2], pp. 66-70.
3. Ritter, J. E., Choi,S. R., Jakus, K, Whalen, P. J., and Rateick, R. G., 1991, "Effect of Microstructure on the Erosion and Impact Damage of Sintered Silicon Nitride," J. Mater. Sci., 26, pp. 5543-5546.
4. Akimune, Y, Katano, Y, and Matoba, K, 1989, "Spherical-Impact Damage and Strength Degradation in Silicon Nitrides for Automobile Turbocharger Rotors," J. Am. Ceram. Soc., 72[8], pp. 1422-1428.
5. Knight, C. G., Swain, M. V., and Chaudhri, M. M., 1977, "Impact of Small Steel Spheres on Glass Surfaces," J. Mater. Sci., 12, pp.1573-1586.

6. Rajendran, A. M., and Kroupa, J. L., 1989, "Impact Design Model for Ceramic Materials," J. Appl. Phys, **66**[8], pp. 3560-3565.
7. Taylor, L. N., Chen, E. P., and Kuszmaul, J. S., 1986 "Microcrack-Induced Damage Accumulation in Brittle Rock under Dynamic Loading," Comp. Meth. Appl. Mech. Eng., **55**, pp. 301-320.
8. Mouginot, R., and Maugis, D., 1985, "Fracture Indentation beneath Flat and Spherical Punches," J. Mater. Sci., **20**, pp. 4354-4376.
9. Evans, A. G., and Wilshaw, T. R., 1977, "Dynamic Solid Particle Damage in Brittle Materials: An Appraisal," J. Mater. Sci., **12**, pp. 97-116.
10. Liaw, B. M., Kobayashi, A. S., and Emery, A. G., 1984, "Theoretical Model of Impact Damage in Structural Ceramics," J. Am. Ceram. Soc., **67**, pp. 544-548.
11. van Roode, M., et al., 2002, "Ceramic Gas Turbine Materials Impact Evaluation," ASME Paper No. GT2002-30505.
12. Richerson, D. W., and Johansen, K. M., 1982, "Ceramic Gas Turbine Engine Demonstration Program," Final Report, DARPA/Navy Contract N00024-76-C-5352, Garrett Report 21-4410.
13. Boyd, G. L., and Kreiner, D. M., 1987, "AGT101/ATTAP Ceramic Technology Development," Proceeding of the Twenty-Fifth Automotive Technology Development Contractors' Coordination Meeting, p.101.
14. van Roode, M., Brentnall, W. D., Smith, K. O., Edwards, B., McClain, J., and Price, J. R., 1997, "Ceramic Stationary Gas Turbine Development – Fourth Annual Summary," ASME Paper No. 97-GT-317.
15. (a) Choi, S. R., Pereira, J. M., Janosik, L. A., and Bhatt, R. T., 2002, "Foreign Object Damage of Two Gas-Turbine Grade Silicon Nitrides at Ambient Temperature," Ceram. Eng. Sci. Proc., **23**[3], pp. 193-202; (b) Choi, S. R., et al., 2004, "Foreign Object Damage in Flexure Bars of Two Gas-Turbine Grade Silicon Nitrides," Mater. Sci. Eng. **A 379**, pp. 411-419.
16. (a) Choi, S. R., Pereira, J. M., Janosik, L. A., and Bhatt, R. T., 2003, "Foreign Object Damage of Two Gas-Turbine Grade Silicon Nitrides in a Thin Disk Configuration," ASME Paper No. GT2003-38544; (b) Choi, S. R., et al., 2004, "Foreign Object Damage in Disks of Gas-Turbine-Grade Silicon Nitrides by Steel Ball Projectiles at Ambient Temperature," J. Mater. Sci., **39**, pp. 6173-6182.
17. (a) Choi, S. R., et al., 2005, "Effect of Projectile Materials on Foreign Object Damage of a Gas-Turbine Grade Silicon Nitride," ASME Paper No. GT2005-68866; (b) Choi, S. R., et al., 2006 "Foreign Object Damage in a Gas-Turbine Grade Silicon Nitride by Spherical Projectiles of Various Materials," NASA TM-2006-214330, NASA Glenn Research Center, Cleveland, OH.
18. Choi, S. R, Bhatt, R. T., Perrira, J. M., and Gyekenyesi, . P., 2004, "Foreign Object Damage Behavior of a SiC/SiC Composite at Ambient and Elevated Temperatures," ASME Paper No. GT2004-53910.
19. Choi, S. R., Alexander, D. J. and Kowalik, R. W., "Foreign Object Damage in an Oxide/Oxide Composite at Ambient Temperature," ASME Paper No. GT2008-50505 (2008); to be presented in ASME Turbo Expo 2008, June 9-13, 2008, Berlin, Germany.
20. Mattoni, M. A., et al., 2005, "Effects of Combustor Rig Exposure on a Porous-Matrix Oxide Composite," J. App. Ceram. Tech., **2**[2], pp.133-140.

21. Simon, R. A., 2005, "Progress in Processing and Performance of Porous-Matrix Oxide/Oxide Composites," *ibid*, **2**[2], pp. 141-149.
22. Peralta and Yoshida, H., 2003, Ceramic Gas Turbine Component Development and Characterization, van Roode, M, Ferber, M. K., and Richerson, D. W., eds., Vol. 2, pp. 665-692, ASME, New York, NY.
23. Cosgriff, L. M., Bhatt, R., Choi, S. R., Fox, D. S., 2005, "Thermographic Characterization of Impact Damage in SiC/SiC Composite Materials," Proc. SPIE, Vol. 5767, pp. 363-372 in Nondestructive Evaluation & Health Monitoring of Aerospace Materials, Composites, and Civil Structure IV.

PROCESSING AND PROPERTIES OF FIBER REINFORCED BARIUM ALUMINOSILICATE
COMPOSITES FOR HIGH TEMPERATURE RADOMES

Richard Cass, Geoffrey Eadon, Paul Wentzel
Advanced Cerametrics, Inc.
Lambertville, New Jersey, USA

ABSTRACT
Advanced Cerametrics, Inc. (ACI) has developed celsian (strontium-doped barium aluminosilicate, SBAS) matrix/celsian fiber ceramic composites that meet the needs for electromagnetic windows on the next generation of hypersonic missiles. The windows have to withstand temperature rises of several thousand degrees in a few seconds, resist impact erosion of Mach 6 raindrops or ice while being electronically transparent to radio-frequencies (RF). Celsian fibers are made using the Viscous Suspension Spinning Process (VSSP), which was developed and patented by ACI. ACI's BAS ceramics have a dielectric constant (K) of about 6.8. A fibrous SBAS composite has K of about 6, which is desirable from an RF standpoint. An additional important feature of the BAS ceramics is their high thermal conductivity and thermal shock resistance. The BAS material is sufficiently strong, but ACI is developing methods to make the BAS stronger and more wear resistant to meet the needs of the next generation of hypersonic missiles. Prototype radomes of BAS-fiber reinforced BAS structures were prepared to characterize the mechanical properties. These radomes are about ¼ the price of reaction-bonded silicon nitride (RBSN) radomes and are much simpler to produce in volume.

INTRODUCTION
Monoclinic barium aluminosilicate or celsian ceramics have generated significant interest in high temperature applications such as electromagnetic (EM) windows, where microwave transparency is required at high frequencies and temperatures. Material properties required for such high temperature applications include low and stable dielectric constant, low dielectric loss, low coefficient of thermal expansion, high thermal shock and erosion resistance, and high use temperature and oxidation resistance. Barium aluminosilicate, $BaO \cdot Al_2O_3 \cdot 2SiO_2$ (BAS) is one of the compounds in the BaO–Al_2O_3–SiO_2 ternary system and has a relatively high melting point of 1760 °C[1]. The materials properties of interest for radome applications are shown in Table I for BAS along with those of fused silica, silicon nitride, Pyroceram, and Cerablak (modified aluminum phosphate) for comparison. Even though fused silica has a very attractive low dielectric constant and low density, it suffers from low strength and maximum use temperature. Pyroceram also has a low maximum use temperature as well as a high coefficient of thermal expansion, which can be detrimental at high-rate temperature changes. Silicon nitride has high strength and relatively high use temperature, but also has a high dielectric constant, which is not desirable for EM window and radome applications. Cerablak has excellent properties for radomes such as low dielectric constant, low CTE, high strength, and high melting point, however, processing of large Cerablak parts has major challenges and it has been mostly used for surface coating. BAS also has excellent properties, however, it suffers from low strength. Increasing the strength of BAS, in the form of a fiber composite, is expected to greatly expand its high temperature applications. BAS exists primarily in three different polymorphs. From 1760 to 1590 °C, the hexagonal polymorph of BAS, hexacelsian, is thermodynamically stable. Below 1590 °C, the monoclinic polymorph, celsian is the thermodynamically stable phase, however, metastable hexacelsian can exist below 1590 °C. At 300 °C, metastable hexacelsian transforms to a body centered cubic orthorhombic phase, a reversible transformation with an associated volume change[2]. In this work, 20 mol.% SrO was added to pure BAS for promoting the formation of monoclinic

$BaO.Al_2O_3.2SiO_2$[3]. BAS is considered a promising matrix material for high temperature structural ceramic composites for propulsion and power systems due to its high melting temperature, and resistance to reduction and oxidation[4]. The monoclinic form of BAS, celsian, has great potential as a next generation radome ceramics due to its very low coefficient of thermal expansion (CTE) (2.5×10^{-6} C^{-1}), relatively high use temperature, and relatively low and thermally stable dielectric constant, ε (6–7), and tan $\delta^{5,6,7}$. However, the mechanical properties of celsian BAS are rather poor with a room temperature fracture strength and toughness of 80 MPa and 1.8 MPa√m, respectively, as reported by Talmy and Haught. In this work, a BAS fiber composite was manufactured and evaluated. The processing techniques and properties are discussed.

EXPERIMENTAL PROCEDURES
Production of SBAS Powder
Barium aluminosilicate powder used in this study was a mixture of 75 wt.% SrO-doped BAS (0.8BaO, 0.2SrO)·Al_2O_3·2SiO_2 (SBAS) and 25 wt.% high temperature eutectic composition (HTE) in the BaO-Al_2O_3-SiO_2 ternary system containing BaO 58, Al_2O_3 10, and SiO_2 32 wt. % with melting temperature of 1320 °C. The addition of the HTE decreases the sintering temperature of the composite and promotes the formation of the monoclinic celsian crystalline phase[8,9]. The composition was calculated based on the starting oxides (Al_2O_3 and SiO_2) and carbonates (BaCO_3 and SrCO_3). The mixture of starting powders was milled and calcined at high temperatures for 1-5 hours. Processing steps for the manufacturing of SBAS are depicted in Figure 1.

Fiber Manufacturing
SBAS reinforcing fibers were fabricated via the ACI's viscous suspension spinning process (VSSP)[10]. In this process, cellulose is digested in a sodium hydroxide aqueous solution in the form of a viscous liquid (viscose) that is mixed with the SBAS powder dispersed in a water slurry (Figure 2a). It should be noted that the lower amount of HTE, compared to the matrix composition, was used in the mixture for fibers to increase their sintering temperature and prevent strong fiber-to-matrix bonding.

Table I. Comparison of physical properties of radome ceramics

Property	Fused Silica	Pyroceram 9606	Silicon Nitride	Cerablak	BAS
Density (g/cm³)	2.2	2.6	3.2	2-2.5	3.37
Toughness (Mpam⁻⁰·⁵)	0.8	2.5	3.5-5.0	-----	1.8
Strength (MPa)	48	300	620-1100	-----	78
Young's Modulus (Gpa)	34-55	110	305	-----	100
Hardness (Gpa)	5	7	15	7	5
Melting Point (°C)	1650	-----	1900	1600	1760
Max. Use Temperature (°C)	1000	1093	1200	-----	<1590
Dielectric Constant	3.3	5.5	8-9	3.3-5	6.5
Dielectric Loss	2x10⁻⁵	3x10⁻⁴	8x10⁻³	-----	3x10⁻⁴
CTE (x10⁻⁶ °C⁻¹)	0.6	5.7	3.0-3.4	2.75	2.3
T. Conductivity (W/m °K)	0.8	3.3	9-30	1-1.5	1.5-1.9

This mix is then pumped through numerous holes in a spinneret into a bath of warm, mild sulfuric acid with a high concentration of a salt. The acid/base reaction coagulates the cellulose into rayon fiber with a high volume percentage of the ceramic powder. The salt dehydrates the water to form the fiber, significantly reducing the diameter and densifying the packing of the ceramic particles. After washing, this fiber is spooled and can be handled as a textile (Figure 2b). The SBAS fibers undergo a binder removal and sintering cycles to reach full density. The fugitive carrier, viscose, is the precursor used for carbon fiber and thus leaves no residual contaminants to lodge in the ceramic grain boundaries after burnout and sintering. Ceramic fibers with diameters as small as 13 μm can be manufactured by VSSP. This process is robust, readily scalable and low cost. Figure 3 shows the microstructure of a fully sintered ceramic fiber.

Composite Manufacturing
    The radome-shaped parts were made by isostatic pressing at 96 MPa using a wet bag isostatic press (National Forge Company, Andover, MA). Radome isostatic pressing tools were designed and built, including a mandrel and a dip tool, to manufacture 16.5 cm diameter green radomes (Figure 4). The fibers with a diameter of 125 μm and length from 0.5-1 mm were uniformly incorporated via a dry milling process into the SBAS powders in the amount of 25 wt.% (Figure 5a). The mixture was isostatically pressed and the composites were sintered in air at 1425 °C for 1 hour. After sintering, the radomes were coated with aluminum phosphate slurry (K=3.3-5) and then refired to increase surface erosion resistance. Figure 5b shows graceful fiber pullout in the fracture surface of the sintered composite.

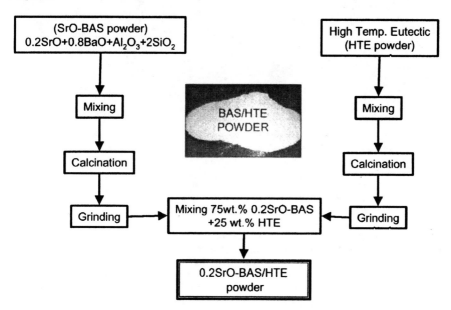

Figure 1. Process flowchart for the production of monoclinic barium aluminosilicate (celsian).

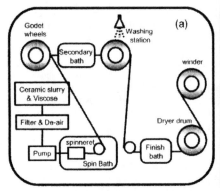

Figure 2. The VSSP fiber manufacturing process: (a) process schematic, and (b) spools of green 0.2SrO-BAS/IITE fiber.

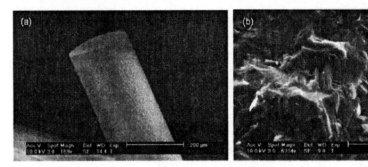

Figure 3. Microstructure of a sintered barium aluminosilicate (BAS) fiber showing: (a) a single fiber, and (b) microstructure of a fractured surface.

Characterization

Mechanical, thermal, and electrical characterizations were conducted at room and elevated temperatures. Thermal conductivity and specific heat were measured at 100, 300, 400 and 500 °C with TPS (Transient Plane Source) and a DSC822e/40 Differential Scanning Calorimeter (DSC, ThermTest Inc., Canada), respectively. The mechanical tests included room temperature bending strength (4-point), thermal shock resistance test, which was conducted by heating sintered parts to 1400 °C and then quenching in room temperature water, and erosion by sand blasting using an in house sand blast machine utilizing a sandblaster supplied with compressed air at 70 psi. The sandblaster forces coarse SiO_2 abrasive through a 3/8" nozzle with an airspeed of ~450 mph. A rain erosion test was performed at Redstone Arsenal for the Naval Surface Warfare Center, Carderock Division (NSWCCD). The impedance was measured using a HP4194 impedance analyzer and used to calculate the dielectric constant.

RESULT AND DISCUSSION

Figure 6a shows the XRD pattern of SBAS/HTE (75/25 wt. %) powder, which indicates the presence of only monoclinic celsian. No hexagonal phase, which is associated with abrupt volume changes at 300 °C in heating or cooling, causing cracks and fractures, was identified. This composites exhibited good thermal shock resistance. No cracks were observed after water quenching from 1400 °C (Figure 6b). The sintered parts demonstrated a dielectric constant of about 6.8 up to 10 MHz, which is sufficiently low for electromagnetic windows (Figure 7). The slight increase in the dielectric constant, above this frequency, is believed to be the machine limitation.

Surface coating of the BAS parts with a thin layer of aluminum phosphate increased their erosion resistance. Backscatter SEM showed a coating thickness of about 250 μm (Figure 8). The coated radomes were subjected to a sand erosion test at 45° and 90° angles and the amount of eroded material was measured by weighing the radomes in every 30 seconds. In this test, the coated samples showed a 40% improvement in SiO_2 erosion test over uncoated samples. These samples have also passed the 35° "Rain Erosion" test at 3.3 Mach speed performed by NAVSEA. Further, the 4-point bend test, shown in Table II, indicates an average strength of above 100 MPa for monolithic radomes without the fiber. The parts with the fibers showed lower strength, which could be due to the defects created when the fiber was added. This observation is still under investigation.

Stable thermal properties are very important in high temperature applications. The thermal conductivity was between 1.5-2 W/mK and was relatively low up to 500 °C (Figure 9a). Further, the specific heat was under 0.6 kJ/kgC and was very stable as the temperature increased (Figure 9b).

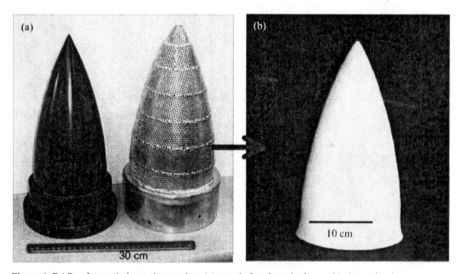

Figure 4. BAS radome via isostatic pressing: (a) mandrel and meshed can, (b) sintered radome.

Figure 5. Barium aluminosilicate fiber composite: (a) powder/fiber (25wt%) mix, and (b) the fracture surface of an isostatically-pressed and sintered fiber composite.

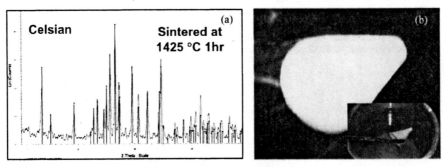

Figure 6. Evaluation of (0.2SrO-BAS)/HTE (75/25wt.%): (a) XRD showing celsian phase, and (b) thermal shock test.

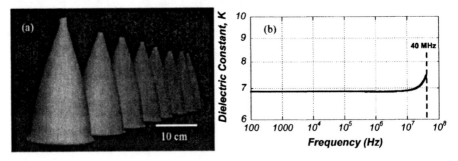

Figure 7. (a) Sintered celsian BAS radomes via isostatic pressing and (b) dielectric constant vs frequency.

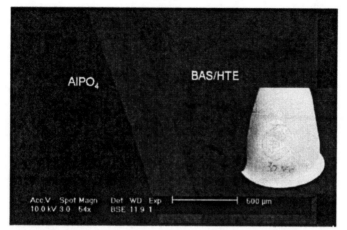

Figure 8. Aluminum phosphate-coated barium aluminosilicate (BAS) showed 30-40% improvement on erosion test. The eroded area during a sand erosion test is circled in the right side photograph.

Table II. Four-point modulus of rupture (MOR) of aluminum phosphate-coated BAS parts.

	Sample No.	Dimension (mm)				Force (N)			Stress (MPa)
		Length	Height	Breadth	Inner-outer loading pin gap	Measured	Offset	Failure	Failure
WITHOUT FIBER	1	49.92	3.04	3.89	5.45	312.2	16	296.2	134.71
	2	50.05	2.99	3.91	5.45	220.0	16	204.0	95.42
	3	49.93	2.99	3.92	5.45	285.1	16	269.1	125.55
	4	49.93	3.03	3.92	5.45	282.2	16	266.2	120.94
	5	50.02	3.07	3.96	5.45	220.0	16	204.0	89.37
	Average								113
25 WT% FIBER	6	50.00	2.98	3.99	5.45	187.0	16	171.0	78.91
	7	49.98	3.00	4.00	5.45	160.1	16	144.0	65.45
	8	50.02	3.04	4.04	5.45	184.9	16	168.9	73.96
	9	50.06	3.00	3.99	5.45	184.9	16	168.9	76.90
	10	49.90	2.96	3.98	5.45	168.6	16	152.6	71.55
	Average								73

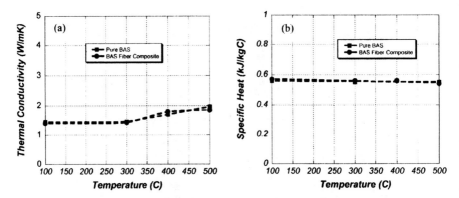

Figure 9. Thermal behavior of SrO-doped BAS and BAS fiber composites: (a) thermal conductivity, and (b) specific heat.

CONCLUSION

Advanced Cerametrics has developed a new high temperature composite using various ratios of 0.2SrO-BAS and a high temperature eutectic (HTE) composition in the BaO-SiO$_2$-Al$_2$O$_3$ ternary system for fiber and matrix. The composite was coated with a variation of aluminum phosphate to increase surface erosion resistance. This material has shown very good mechanical, dielectric, and thermal properties and has successfully passed a rain erosion test at a speed of 3.3 Mach. The BAS composite has great potential for high temperature applications such as electromagnetic windows, reentry vehicle leading edges, catalytic converters, thermal insulation, and high temperature substrates for electronics. The process emerging from this work produces a competitive material to RBSN at a small fraction of the cost with considerably greater ease of manufacturing.

ACKNOWLEDGEMENT

This work was supported by the Office of Naval Research (ONR). ACI is currently completing a very successful Phase II SBIR with Navsea at Carderock for the development of hypersonic radome materials (contract # N00167-05-0008, TPOC Curtis Martin and Dr. Inna Talmy) and a Phase I SBIR with Navair at China Lake to develop hypersonic window materials (contract # N0014M0326, TPOC Pam Overfelt).

REFERENCES

[1]. H.C. Lin and W.R. Foster, "Studies in The System BaO-Al$_2$O$_3$, I. The Polymorphism of Celsian," *The American Mineralogist*, Vol. 53, P134, 1968.

[2]. B.Yoshiki and K Matsumoto, "High-temperature modification of barium feldspar," *J. Ceram. Soc.*, Vol. 34, PP 283-286, 1951.

[3] I.G. Talmy and D.A. Haught, "Ceramic Material" *United States Patent*, no. 5,642,868, 1997.

[4]. J.J. Buzniak, K.P.D. Lagerlof and N. P. Bansal, "Hot Pressing and High Temperature Mechanical Properties of BaAl$_2$Si$_2$O$_8$ (BAS) and SrAl$_2$Si$_2$O$_8$ (SAS)," *Ceramic Transactions, Advances in Ceramic*

Matrix Composites II, Edited by J. P. Singh and N. P. Bansal, American Ceramic Society, Westerville, OH, Vol. 38, PP. 789–801, 1993.

[5]. I.G. Talmy and D.A. Haught, "Ceramics in the System $BaOAl_2O_32SiO_2$–$SrOAl_2O_32SiO_2$ as Candidates for Radomes," *Technical Report*, Contract N60921-8-R-0200, Naval Surface Warfare Center, 1989.

[6]. I.G. Talmy and D.A. Haught, "Celsian-Based ($BaO \cdot Al_2O_3 \cdot SiO_2$) Ceramics as Candidates for Radomes," NASA Conference Publication No. 3097, *14th Conference on Composite Materials and Structures, Part 1*. National Aeronautics and Space Administration, Washington, DC, PP. 227–38 1990.

[7]. J.M. Wright, J.F. Meyers and E.E. Ritchie, "Development of Celsian ($BaO \cdot Al_2O_3 \cdot SiO_2$) Ceramics for Advanced Radome Applications," NASA Conference Publication No. 3175, *16th Conference on Metal Matrix, Carbon, and Ceramic Matrix Composites, part 2*, National Aeronautics and Space Administration, Washington, DC, PP. 845–54, 1992.

[8] I.G. Talmy and D.A. Haught, "Method for Preparing Monoclinic $BaO \cdot Al_2O_3 \cdot 2SiO_2$," *United States Patent*, No. 5,695,725, 1997.

[9] I.G. Talmy and J.A. Zaykoski, "Sintering Aids for Producing $BaO.Al_2O_3.2SiO_2$ and $SrO.Al_2O_3.2SiO_2$ Ceramic Materials," *United States Patent*, No. 5,641,440, 1997.

[10] R.B. Cass, "Fabrication of Continues Ceramic Fiber by the Viscous Suspension Spinning Process," *Ceramic Bulletin*, Vol. 70, No. 3, PP. 424-429, 1991.

Author Index

Printed in the United Kingdom by
Lightning Source UK Ltd., Milton Keynes
142418UK00002B/90/P